THE SPIRIT
OF
LONDON'S RIVER

THE SPIRIT
OF
LONDON'S RIVER

— Memoirs of the Thames Waterfront

L. M. Bates

GRESHAM BOOKS

In association with the Europa Publications

GRESHAM BOOKS
Unwin Brothers Ltd,
The Gresham Press,
Old Woking,
Surrey
England

First published 1980

ISBN 0 905418 43 3

Set in 11/12pt Janson and printed
by Unwin Brothers Ltd, The Gresham Press,
Old Woking, Surrey, England.

Author's Note

Thamesmen, afloat and ashore, who have helped me are too many to mention individually, and I can only express general gratitude to all those who have checked facts for me about far-off times.

I must, however, particularly record my appreciation of the great pains taken by my old friend and former colleague, Bramwell Evans, to keep me abreast of the changing tidal scene, and of the help of a later friend, Graham Avory.

Again, in the matter of illustrations, while able to give only general thanks to the firms, institutions and River people who have so kindly supplied suitable photographs, I must mark the special help given me by two more friends and former colleagues, Stanley Rawlings and Roy Johnson.

Here and there I have reworked material used in earlier books of mine, long since out of print, and I have also drawn many of my facts from articles I contributed to various journals, particularly the international shipping daily *Lloyd's List and Shipping Gazette* and (then monthly, now quarterly) *The Port of London*.

Opinions expressed in this work are my own and do not necessarily reflect official port policies.

to J. G.

Contents

LIST OF ILLUSTRATIONS

Introduction

THE TIDAL THAMES downstream of London Bridge has, like so much of the Greater London scene, inspired a fair number of books. Most of these have been topographical; a few historical. But only an author who has lived on intimate terms with the tideway* could portray those elusive qualities which make it unique among waterways.

Physically insignificant by world standards, the Thames has a great port, but so have other British rivers. It has all the modern technology with which such terminals handle cargo, but again so have most commercial havens. It has a history going back nearly two thousand years, like some of those other rivers and harbours. But none other blends past and present so successfully into its working life, merging tradition with technology, great cargo installations with romantic survivals of more picturesque days, computer programming with ancient professions where skilled eyes and hands alone count.

Such an intangible atmosphere is perhaps more easily discerned by the artist, and it has been captured in paint by members of the Wapping Group whose annual exhibitions of Thames paintings in London's Royal Exchange have been aptly named the Spirit of London River.

Fortunate in living on very intimate terms with the men, ships and craft of the old river, I have based much of my narrative on personal experience with the Port of London Authority - at its docks, in its former headquarters on Tower Hill, afloat along the fairway in peace and (with the Royal Navy) in war - as well as taking some of the distant past as a backcloth to the present.

Fresh from school at the fag end of the First World War, I found dockland a very tough place indeed. When the Port of London Authority was constituted in 1908, a much-needed programme of new plant, offices and equipment was contemplated. But the outbreak of war had frozen much of the work in hand, and the Port still had the same ramshackle equipment which the more impecunious of the former private dock companies had handed over to the PLA.

The Authority, besides absorbing the administrative staffs of the dock

---

* Joseph Conrad and H. M. Tomlinson come to mind as authentic voices of the past.

companies (several of them semi-literate labourers promoted as a reward for blacklegging in some far-back strike) had begun recruiting from public and grammar schools. But the war had snatched away the young men, leaving the docks staffed by an odd assortment. All the older men wore their bowler hats (then status symbols for clerks, foremen and tugmasters) in the office. All shook hands on arrival and departure; some even when they went out to lunch. Smoking then as now was forbidden at the docks and most of the old hands took snuff. At lunchtime they adjourned to their beer club in the basement; common dockside pubs were not for the likes of them. As one put it: 'You can't expect us blee'n officers to drink with the men; it'd undermine discipline.' Also by buying beer in the cask they saved on the price.

The warehouses and dock offices were infested by large bold rats and war was waged on them by professional rat catchers who visited their trap lines daily. They were helped by hordes of half-wild cats which were officially recognised by a ration of meat and milk.

Most of the cats have long since gone but in the Authority's former museum was a gruesome relic of those days - the mummified body of a cat with its claws sunk in the neck of a mummified rat, found wedged in one of a labyrinth of runs when one of the old dock offices was demolished. Some of those cats were great travellers, nonchalantly making a voyage or two to the Far East or other distant places, eventually returning to their dockland duties of rat catching and breeding.

The office where I began work had a phallus-like instrument known respectfully as **THE TELEPHONE**. But it was too new to be regarded as trustworthy and, although by then the telephone network had spread over most of London, certain formal communications to the headquarters in the City were primly tapped out on morse keys by two elderly lady telegraphists.

Clients at the dock offices were mainly carters presenting documents of title before collecting cargo. Most of dockland's traffic was then horse drawn, and every morning a great concourse of carts, vans and wagons streamed through the dock gates. These visitors brought with them a rich smell of stable manure, sweat and oiled leather. We were envious of a neighbouring office where carters awaiting delivery orders to be passed had, over the years, completely covered the walls with graffiti - all highly obscene.

In the warehouses and out on the quays were the dockers, in those days the neglected waifs of industry. They almost literally fought for work at the morning 'call on'. Many failed. And when ships were delayed, none of them earned enough to meet the family's basic needs - food, warmth, shelter. There was then no attendance pay nor even the later microscopic unemployment 'dole'. So a bad week for the docker meant the Sunday suit and the watch going into the pawnshop until ships came in again.

In those days there were few mechanical aids at the docks, except for hydraulic quay cranes and grain elevators. This meant that practically all the boxes, cases, sacks, casks and so on were handled manually. There had been little change since 1851 when Henry Mayhew wrote: 'This class of labour is as unskilled as the power of the hurricane.' Most of these port workers were literally clothed in rags until demobilisation got under way when cast-off khaki became almost dockers' uniform.

Most dockers then spoke a common tongue, almost unintelligible to the outsider. Oaths and obscenities were used without rancour or emphasis. It was then the fashion to interpolate the lurid adjective between the syllables of the noun so qualified. Thus, Yokohama would become something like Yokoblee'nhama, and the range was unlimited.

Dockland similes were borrowed from the imaginative metaphor of the sea. I am rather too old-fashioned and perhaps too prudish to take kindly to modern literary permissiveness, so I must bowdlerise. A 'dry' cooper, asked if he had finished securing the head of a hogshead of leaf tobacco, would probably reply: 'Tight as a nun's private parts'. Or, commenting on the ease with which the same head had been removed, might say: 'Came off like a Jew's appendage'. There would be no snigger; it was the right and natural way to describe the work.

In my time, too, there were still traces at the docks of the wave of gibberish known as backslang which swept through London's East End behind the better-known rhyming slang. Even today, night work along the Thames is sometimes called 'tidgin' - backslang for night - although few who use the term know its origin. Then, as now, a load falling out of a crane sling was a 'Greenacre', so named after a murderer at whose execution the gallows collapsed, taking both him and the hangman through the trap.

During my early days at the docks the Port of London Authority's police marched on their feet or, at best, rode bicycles. They had no Z cars, CID, radio-telephony, motor-cycle patrols and other refinements which today help to keep London's enclosed docks comparatively free from crime. At the end of the First World War, their greatest problem was pilferage of cargo.

The dock warehouses then held what was one of the world's largest and most valuable concentrations of easily portable merchandise. To men living on the edge of privation, the urge to pocket some of the goodies which they could never afford to buy was often irresistible. I once saw an apparently pitiful little humpback stopped at the dock gate where the policeman on duty removed the hump from the back of his coat - it was a frozen leg of lamb stolen from the cold store.

Shipments of liquor were always tempting. One docker was charged with stealing a bottle of whisky. He assured the magistrate that the bottle had contained medicine when he hung up his coat in the ship's hold. The magistrate commented that the only comparable case on record, the changing of water into wine, had been regarded as a miracle. 'If it could happen once, it could happen again,' suggested the docker hopefully. 'Two months,' replied the magistrate.

The state of the nation at the end of the First World War was reflected in the ships using London's docks. A few fine vessels had survived, but many others were floating junk shops, only kept in service until they could be replaced. They were unlovely with their batteries of no-nonsense funnels as upright and thin as telegraph poles and nearly as tall, their harsh and angular superstructures and rust-streaked hulls. Many of the forecastles where the crews lived were vermin-infested slums.

These crews varied from Scandinavians, Somalis and West Indians to Japanese and other exotic races. Chinese seamen had the delightful custom of placing little dishes of delicacies in their engine rooms to propitiate the invisible devils who drove the ships.

Lascars, as Indian seamen were then called, flapped along in their flimsy cottons during winter as well as summer. Two and sometimes three generations of a family might be found serving in one vessel, with perhaps many relations in sister ships of the same fleet. A festival of Muslim seamen, which appears to have died out in dockland, involved a long and noisy religious procession held during the first month of the Muslim year. It commemorated the deaths of the Prophet's nephews, Hassan and Hussein, and was known to dockers by its Anglo-Indian slang equivalent of 'Hobson-Jobson' Day. Ethereal-looking Goanese, Roman Catholics all, made up a large proportion of the ships' stewards.

With the 1920s, changes began. The young men came back from the war bringing with them the twentieth century. The lady telegraphists went into genteel retirement. Telephones proliferated. Ernest Bevin became the 'Dockers' KC', winning for them the then stupendous wage of 16s (old money) a day for a forty-four hour week. The four-letter word which had done duty in the trenches as noun, adjective and adverb started to erode the rich obscenity of dockland speech. Lorries began to replace the horse-drawn vehicles and the PLA hastened to make up for wartime delays in its immense task of modernising the Thames and its docks.

I had arrived just in time to witness the end of an epoch, but the changes made little outward difference to the normal tenor of dock working. And so it continued until after the Second World War. Then, little more than a decade ago, the Port was suddenly completely transformed, and the pace of change is still accelerating. Steamship lines, whose names have been a part of the London scene for a century or more, have declined or departed; new companies owning ships of outlandish design with outlandish names have taken their place. Trades, so ancient that they had become merged in tideway mythology, have disappeared; methods of packing, transporting and distributing cargo have been revolutionised. Docks and wharves which were milestones in London's history have abruptly gone out of business. And many local skills have been lost.

The apex of the commercial port has already moved downstream to Tilbury in Essex. Near where the trained bands and London apprentices had gathered under Elizabeth I to meet the threat of the Spanish Armada, container ships and bulk carriers are served by complicated giant handling machines driven by men whose fathers relied mostly on their own muscles. But horizons continue to broaden and the future may lie still further seaward at Maplin whose sands have seen Roman galleys, Viking longships, de Ruyter's raiding fleet and German aerial minelayers.

So perhaps this nostalgic record of what a very ordinary servant of the Port has read, seen and heard, from the days when commercial windships still brought grain and timber to London, succeeded today by Jumbo-sized carriers, may help to keep alive memories of much that my generation thought was eternal but which vanished almost overnight. And out of my half-century of rubbing shoulders with so much historical, topographical and sociological material may emerge something of that unique Spirit of London's River. Not for nothing has the tidal Thames been called our island line of life and fate.

SECTION ONE

DOCKLAND AND TIDEWAY BETWEEN THE WARS

Title illustration:

The famous old clipper *Cutty Sark* in her permanent berth at Greenwich. *(Courtesy Brian Tremain AIIP, FRPS, National Maritime Museum.)*

chapter one

From London Bridge

1

THE UPPER POOL is that part of the tidal Thames between London and Tower Bridges. Here was the starting point of my personal story of the lower tideway and its banks on its journey down to the sea; the old-time sailors' London River.

The River of London (the upper tideway), west of the Pool to the Port's landward limit of Teddington, with its network of bridges barring the way to big ships, has its rich history and bankside splendours, yet these did not compensate me for the absence of that rumbustious workaday life which teemed along the lower river reaches.*

My recollections are not a step by step survey of the riverside though, for clarity, they follow the river downstream, but a selection of my impressions and encounters gathered over nearly fifty years.

John Rennie's nineteenth-century London Bridge always seemed to me something of a bridge of sighs for those many generations of commuters who twice daily trudged its windy length. Below them, downstream, they could see brave symbols of a freer life - flags, masts, funnel smoke, strange cargoes and ships leaving on the tide. This is where the Port and City began some two thousand years ago, but the modern Port of London had virtually moved away downstream even before the Second World War.

Now Rennie's bridge (one of whose predecessors was that more

---

* A reach is a straight part in a navigable river and serves as a convenient method of geographical demarcation.

picturesque but highly inflammable structure, cluttered with houses, shops and the busy, self-contained lives of its inhabitants) has been sold to America. The new London Bridge, designed for the motor age, has now become an accepted part of the City's crowded scene.*

It was from an arch of Rennie's bridge that I first saw a truss of straw hanging; the official warning of some temporary obstruction (such as a workman's cradle) which reduced clearance for passing vessels. At first I was inclined to smile at so homespun a method but I could not think of a better substitute. The warning had to be clearly visible on the darkest day; sufficiently flimsy not to damage topmasts or superstructures striking it; sufficiently flexible to remain undamaged by any such impact; and unlikely to disintegrate during bad weather. In search of historical clues about its origins, I consulted the City of London Guildhall Library where research staff came to the conclusion that the custom might go back to Roman times.

Tower Bridge - the last bridge over the Thames on its way to the sea - is best seen from the head of Limehouse Reach when bound upstream in a clear winter dusk. Then the lights of road traffic moving across the bridge borrow from the pageantry of older and more colourful days and become a glittering cavalcade of surpassing beauty.

I never tired of watching when inward-bound sea-going ships had signalled to Cherry Garden Pier (some half mile downstream) by hoisting a pennant and black shape, accompanied by one long and three short blasts on the whistle (Morse Code B). On receiving the message from the pier, the Bridge Master had the road traffic halted and put simple, efficient and practically foolproof machinery into operation. The engine room, underneath the south approach road, was where powerful steam pumps fed hydraulic accumulators.† As for the exterior of the bridge and the sight of its two arms opening in a benedictory salute to ships, no description is needed; this is still one of the world's famous bridges. (It used to open an average of fourteen times a day; now as infrequently as twice in one week.)

---

* Opened by HM The Queen in March 1973.
† The bridge is now operated by electro-hydraulic machinery. One pumping engine, no longer required, was presented to an industrial museum in 1975. Another is to form part of a small Tower Bridge Museum.

Plate 1:

The Howland Great Wet Dock, Rotherhithe, circa 1700. *(From a contemporary print in the collection of the Port of London Authority.)* This was London's first enclosed wet dock. It was later converted into the Greenland Dock, part of the Surrey Commercial group.

Plate 2:

A late stage in the construction of the St Katharine Dock at Wapping, opened in 1828. *(From a contemporary print in the collection of the Port of London Authority.)*

2

Lining the south bank between the two bridges were variously named wharves grouped under Hay's Wharf, Ltd.* The 300-year-old history of this undertaking is interwoven with the ancient chronicles of Southwark, with old London Bridge, with the story of the Upper Pool and, in more recent times, with the London provision trade centred on Tooley Street. China clippers formerly berthed at the Hay's Wharf complex and Shackleton's *Quest* sailed from there on his tragic last voyage to the South Pole. In the thirties, Russian ships regularly discharged there and some of them reflected the earlier insecurity of that country's masters. Such vessels were named after currently prominent revolutionaries. A fall from office usually meant the renaming of the ship.

On the north bank below London Bridge is a half mile of riverside which records, perhaps in more detail than anywhere else, something of the continuity of London's great measure of time past. Close to the bridge is Billingsgate, the best-known fish market in this country, and enshrining in its name that of Belinus, the Roman engineer who is reputed to have built Stane Street.

Up to the late thirties, two Dutch eel schuyts swung to a mooring here, supplied with eels from the Netherlands. The story was that during the Great Plague of 1665 the eel schuyts had braved the danger and continued to run supplies to London, after which a grateful king had granted free mooring in perpetuity.

Then in 1938 the Dutch sold these two schuyts to a British owner who kept one at the Billingsgate mooring, buying eels ashore in the market to smoke on board for resale. When it was decided to close the business, an offer was made to sell the mooring to the PLA. But a thorough search revealed no trace of any grant, so it appears that the crafty Dutchmen had occupied the mooring for nearly three centuries simply by squatters' right.

Downstream of Billingsgate is the Custom House; there has been a

---

* Closed and due to be redeveloped.

centre for the collection of revenue on or near this site for some six hundred years.

Still further downstream stood Galley, Brewer's and Chester Quays (all destroyed in the war and now covered by Three Quays, offices built by the General Steam Navigation Co Ltd). Galley Quay was where the medieval wine fleets from Venice and Genoa discharged, though the area has connections with the importation of wine going back perhaps to the Roman occupation. It was fitting that the General Steam should settle on the site, for this company was for many years one of the principal carriers of this ancient trade.

Nowhere else is so much of the story of Thames-side London made so plain as at Berkyngechirche, All Hallows by the Tower (once in the possession of the former Benedictine Abbey of Barking).

History here begins with the well-preserved Roman pavement in the Undercroft, part of the Roman house occupying the site before the church was built in the seventh century. Among the relics found there is a toy stone lamp with which some Roman child once played. Perhaps, like the pathetic toys on the heaps of rubble after the winter of 1940-1, it was lost in the terror of London's first recorded 'blitz', for the foundations of the Undercroft show a layer of burnt earth and charred debris, proof of Boudicca's vengeful raid of AD 61.

Milestones to the centuries include the Saxon arch, claimed to be by far the oldest in London, and the Saxon stone cross which probably stood on Tower Hill long before the Normans came and which was uncovered by bomb damage; the twelfth-century altar at which Richard Coeur de Lion worshipped; the complete church registers since 1558; the tower built in Cromwell's time, up which Pepys hastened to watch the Great Fire of London; the Grinling Gibbons font cover; the Toc H lamp, first lit by the then Prince of Wales; and so on.

Some treasures were lost when the church was bombed in 1940. By the efforts of the 38th Vicar, the Reverend Philip Thomas Byard Clayton, CH, MC, ('Tubby' Clayton), and help from Toc H members all over the world, the church was rebuilt.

Neighbouring All Hallows is the Tower of London. In the shingle bank below its riverside promenade (the bank now hidden under the sand of an artificial beach for children) were found Neolithic arrow heads; and Stepney mums and their young, at summertime low water, enjoyed until recently make-believe holidays with deck chairs, picnic lunches, spades, pails and paddling, where early Thames-siders once hunted and fought for the prized river ford. Behind the beach is the bricked-up arch of Traitors' Gate, with its ghosts of those who passed through that grim portal; names too well known to need another roll call.

Bringing the narrative up to our own time, the way to the beach was by a ship's accommodation ladder, peacetime equipment of the former P & O liner *Rawalpindi* and inscribed with the epic account of her last gallant action against two German battleships in 1939. I have asked more than one grubby urchin on the beach what he knew of the story and was pleased to find that, in spite of the noisy Anglophobes in our midst, it was still taught with pride. Alas! Time and weather made the ladder unsafe and beyond repair, and it was recently removed.

chapter two

Wapping and the Upper Docks

1

AT WAPPING ON the north bank, a little below Tower Bridge, was a rambling series of lagoons, leading one from the other; the London Dock.* Neighbouring it upstream, but not physically joined, were the St Katharine Docks.† The former was opened in 1805; the St Katharine Docks followed in 1828.

Two of the entrance locks of the London Dock had been permanently closed, and shipping and craft entered or left by the New Shadwell Entrance, leading from the Shadwell Basin into the Lower Pool.

The comparatively narrow internal dock roads were at the bottom of canyons created by row after row of towering fortress-like warehouses, built to last. Designed by D. A. Alexander, they reflected his other work on lighthouses and jails. An imaginative touch was the skilful display of an occasional ammonite left by the masons in the huge stone plinths.

The complex of basins was used by coasters, small ships in the Continental trade and numerous dumb barges, for the London Dock was the Port's main storage centre. Thus the great bulk of the treasures behind these impregnable walls had been barged up from larger ships that discharged in the deeper docks downstream. In those days, this warehousing trade was of the utmost consequence to the Port.

---

* Closed in 1969.
† St Katharine Docks, closed in 1968, have now been developed as a yacht haven.

Wool, wines and spirits, drugs, iodine, gums, ivory, mercury, rubber, shells, spices, tortoiseshell, essential oils, perfumes, coffee, cocoa, dried fruits and canned goods – these and other commodities had their own special fastnesses where the warehouse staff included samplers, graders, blenders, sorters, bulkers and other specialists. These great caravanserais were bitterly cold in winter and cool in summer, and the low temperatures seemed to add to the distinctive smell, in some cases the stink, of each one.

On their bare stone floors, commerce was made manifest; not the commerce of statistics, accounts and market reports, but the real grass roots of international trade, laid out as one imagined merchants displayed their wares in Babylon or Ur. Here were samples of the world's real wealth produced, not by financiers or brokers, but by the skills and patience of men who, in some cases, could have had no conception of the society which would consume them. Here men spoke of far-flung places, exotic, sonorous or even sinister sounding, as if they were merely two stations down the line.

The wrappers and containers in themselves told a story – perhaps a story of ingenious primitives using what nature had provided in the way of fibres or canes or skins to send their stuffs to market; or a story of highly sophisticated exporters who used rare technological skills to ensure the safe carriage of their merchandise.

When the warehouse staff, usually men of little book learning, spoke of the particular commodities in their care, they became strangely coherent, boldly expressing opinions backed by the knowledge accumulated after a life-time's experience. The origins of the unique processing services for merchants and traders by these men are to be found in two fifteenth-century charters giving the City of London Corporation the right to warehouse and examine goods brought to London. While certain privileges, such as the Corporation's control of the City's wholesale fruit, fish and meat markets, still continue, those rights of handling and warehousing certain cargoes were transferred at the beginning of the nineteenth century to the dock companies, and then in turn, although no longer a monopoly, to the Port of London Authority.

London had for long been the world's wool market and at the London Dock were some thirty acres of storage space to hold about 200,000 bales. The wool came mainly from Australia, New Zealand, South Africa and South America.

About eight acres on the top floors of the wool warehouses were set aside for displaying representative bales before the six auction sales held each year at the Wool Exchange in the City of London. These bales were cut open and prospective buyers from all parts of the world used to examine the quality (for which the true north light of these top floor windows was essential) in readiness to make their bids.

Here was much talk of fleeces, slipe wool (taken from a dead animal), scoured wool (with the natural lanoline extracted) and greasy wool (in its natural state).

Similarly, London was the world market for ivory. Before an auction sale, the Ivory Floor would be covered with rows of tusks laid out for inspection. The tusks were rarely hunters' trophies but mostly 'found' ivory, often from long-buried hoards of tribal chieftains. An occasional tusk splintered and cracked by a bullet usually indicated a rogue elephant slaughtered by game wardens.

The sorting foreman would use a torch to look down the hollow part of each tusk; sometimes this would show the ravages of decay, while grooves on the outside would indicate how the poor beast, agonised by six feet or more of toothache, had tried to ease the pain by sawing it back and forth across the branches of trees. Another tusk might show the marks of crocodile teeth, suggesting the lonely tragedy of an elephant trapped in a swamp.

Perhaps the greatest attractions to visitors were the huge mammoth tusks, taken from animals preserved intact in the frozen soil of Siberia and shipped on a commercial scale to the London Ivory Market. In those days, before carbon dating was used to determine the approximate age of such finds, the heyday of the mammoth was a matter for varied speculation.

Here, too, would be large numbers of rhinoceros horns, nearly all destined for the Far Eastern trade in aphrodisiacs.

Beneath the warehouses and dock roads were about twenty-five acres of vaults, some of whose ceilings were beautifully fashioned in the form of a series of mushroom caps, supported on octagonal stone pillars. Here were stored nearly eighty thousand casks of wines and spirits. Darkness and the slow evaporation of wine bred the characteristic clusters of fungus which hung from the ceilings.

A steady temperature of 60°F, maintained by means of gas jets and

Plate 3:

The Upper Pool from London Bridge in 1841. Although by then some of the enclosed docks further downstream had been built, the Pool was still the virtual heart of the Port of London. Today, the Port is mainly concentrated at Tilbury in Essex. *(From a lithograph by William Parrott in the collection of the Port of London Authority.)*

Plate 4:

Before the war, the last of the deep-sea windships used to bring grain to this country in the so-called 'Grain Race'. This picture shows the *Winterhude*, a frequent visitor to London, in the Millwall Dock. *Photo: Stanley.*

ventilators, was essential for the slow maturing of the wines. The vault floors were thick with sawdust which deadened footsteps; often the only noise, perhaps away in the gloom out of sight, was the tapping of coopers' mallets. These men, on the lookout for leaks, made regular rounds, not unlike railway wheel tappers at a terminus, accurately judging the volume of each cask's contents by the pitch of sound from two or three sharp blows with their 'floggers'.

Coopers were then the aristocrats of dock labour, highly skilled after long apprenticeship. They were proud of their traditions and used special tools, some of them not much altered since the days when man first realised that a rounded container offered less resistance to inertia than a box.

Identification marks and numbers were indelibly scored on the head of each cask with a 'scribing iron'. This resembled a large old-fashioned tin opener, with a point and a blade; letters and numerals were cut by downward strokes and curves almost as quickly as writing with a pen. Such tools had been used for centuries to mark casks and other types of cargo containers.

Bomb damage to 'A' Shed on the East Quay, London Dock, revealed doodles which had been cut into the supporting beams by scribing irons more than a hundred years ago. Most of the inscriptions were merely names and were dated 1806. But one doodler, more enterprising than his fellows, recorded: 'Lord Nelson Dyed [*sic*] in Victory October 21st 1805'.

It was after the First World War that coopers began to use battery hand lamps. Before that, they had oil-fired torches, as wine is not inflammable. (The latter could not be used, of course, where brandy and other spirits were stored.) The old torches, some inscribed with the names of visiting VIPs, were still preserved.

Such visitors were usually taken to the Crescent Vault where they would be met by the dock superintendent who followed a set ritual. First he would assure them that he had the permission of the wine trade to give them a taste. And, as evidence, he would produce a tattered 'tasting order'. Then the vault keeper would select a cask, and one of his coopers would give two or three shrewd blows with his flogger on each side of the 'shive' (bung) to make it jump out. A valinche, a long metal tube like a laboratory pipette, would be plunged into the cask and withdrawn full of wine, and the glasses charged.

In the 1930s, I was helping to entertain here a party of Japanese royalties when language difficulties caused me to be mistaken for the owner of the wine. The guests bowed, hissed politely and addressed me as Mister Sandeman for the rest of their visit.

2

The St Katharine Docks, smallest of those taken over by the PLA, consisted of the Eastern and Western Docks, joined by a tidal basin and with an entrance lock into the Upper Pool, upstream of London Dock.

Like the latter, they were used by coasters, short-sea traders and many barges, for here was another of the Port's storage centres. The six-floor warehouses, functional but also distinctive, were the work of Thomas Telford, the engineer, and Philip Hardwick, the architect, who introduced a classical theme into the design by supporting them on cast-iron columns along the quays.

Telford and Hardwick were also responsible for the first St Katharine Dock House, just outside the dock wall, opposite the Royal Mint.* In the nineteenth century it had contained the Board of Trade offices where examinations for Merchant Service masters and mates were held. Joseph Conrad took all his 'tickets' there and in *A Personal Record* has left an account of the anguish of these trials, on which a Merchant Navy officer's career wholly depended.

Also in the building, handy for the sordid temptations of old Ratcliff Highway, had been the shipping office where crews were signed off or on their vessels. Known colloquially as the 'chain locker', it is described in Conrad's novel *Chance*.

I always regarded the St Katharine Docks as something of a home for lost causes, since I saw there the fitting out and departure of many single-handed small-boat voyages. Some of them were desperate ventures in poverty-stricken ramshackle craft, whose owners sometimes sailed leaving a pile of unpaid bills with ship chandlers and outfitters. And those were the voyagers who usually gave up when they met the chops of the Channel. But these docks were also the starting point for a number of well-found round-the-world-yachts and scientific expeditions.

---

* Moved to Llantrisant, Glamorgan, in 1971.

3

Closely associated with the London and St Katharine Docks as a Port storage centre were the vast, rambling PLA warehouses* in Cutler Street, behind Houndsditch in the City. They were built in 1782 for the Honorouble East India Company and had been inherited, with the East India Docks, by the PLA.

Here was a treasure house of strange and beautiful things, reflected in the visitors' book whose galaxy of royal signatures read something like an extract from the *Almanach de Gotha*. It also included the names of such famous Prime Ministers as Gladstone and Balfour; and those of the actress Sarah Bernhardt, and the late Poet Laureate, John Masefield, who had recorded his appreciation in verse.

There were valuable consignments of tortoiseshell, some pieces gouged by the teeth of predators; luscious scents from drugs and spices; lumps of foul-smelling ambergris from whales' bellies; equally foul-smelling extrusions of the civet cat, arriving in such small, highly priced quantities that it was exported from Abyssinia - or, more fashionably today, Ethiopia - in plugged buffalo horns (the last two commodities used as fixatives in expensive perfumes); and aloes, poured in liquid form into monkey skins and allowed to harden into gruesome, corpselike shapes.

There were rooms reserved for the choicest of cigars; others for Chinese and Japanese porcelain, bronzes, beaten brass, oriental pictures, ancient manuscripts - all on display to traders; stalls of ostrich feathers, sorted and graded ready for the six annual auction sales (before the First World War this trade ran into some £3 millions a year). And before importation was prohibited, there was the showy plumage of tropical birds. Other colour was provided by priceless silks, both raw and manufactured.

Best of all were the acres of rooms holding carpets and rugs from India, Turkey, Persia and China. Some were unique, with a fairy-tale quality; others beautiful in design or colour. Occasionally a newly opened bale of carpets let fall camel droppings and desert sand.

After the Second World War, an important trade grew up in these warehouses - the bottling and binning of vintage wines.

---

* Sold for development in 1973.

At the other PLA town warehouses in Commercial Road, East London, tea from India and Ceylon was sampled and blended. The PLA also owned cold stores at the Smithfield Meat Market.

4

Outside the main road gates of the London Dock at Wapping was The Highway (formerly Ratcliff Highway) which, after the First World War, no longer saw Kipling's seamen 'drunk and raising Cain'. But very old dockers who had known the dock basins full of sailing ships (including Conrad's *Torrens*) could remember when unconscious sailors were robbed, stripped naked and thrown into the Highway covered only by a sheet or two of newspaper. Paddy's Goose, the notorious dockside pub which took its name from an Irish owner and a stone swan on the roof, still flanked the private hall where obscene exhibitions attracted Victorian fashionables.

Both river banks, flanking the Upper and Lower Pool, were especially rich in old watermen's stairs; some disappeared in the 'blitz'; some have been or are being absorbed in new development; others still remain. Many of their names - such as Prince's Stairs, Alderman Stairs, Cherry Garden Stairs, Fountain Stairs, Pelican Stairs, Pickle Herring Stairs - were colourful but inconsequential punctuations to a mainly squalid and utilitarian background.

Between St Katharine Docks and the river was St Katharine's Way, in my time a dark canyon between cliffs of warehouses. And here was South Devon Wharf,* something of a landmark for all good Thamesmen. It was once the home of W. W. Jacobs when his father was manager there. A withdrawn youth, he later drew the characters in his stories from the Wapping sailortown about him.

St Katharine's Way led into Wapping High Street, squeezed between the river and London Dock. Wapping Police Station, headquarters of the Thames Division, Metropolitan Police, is on the river bank off the High Street. Nearby, on the site of the former Morocco and Eagle

---

* Disappeared under redevelopment of this area.

Sufferance Wharves, is their new ultra-modern boatyard. Quite separate from the PLA's own dock force, the River Police patrol the Thames from Dagenham to Teddington Lock. Relations between this force and tideway workers were still guarded in my early days, but growing more friendly. The great wave of river piracy in previous centuries had ebbed away, leaving only a few ripples, and the police themselves were beginning to claim that the tidal river was among the more law-abiding areas of London.

Nevertheless, a hard-up waterman who 'drudged' coal from the river bed at a coaling berth was still liable to be ordered by the police to turn over the pile of salvaged coal in his skiff. If he had stolen it from a barge, merely throwing water over it to make good his story of 'drudging', the bottom layer would have been dry.

The River Police had their own macabre lore, particularly about the places where bodies of suicides, the sport of tidal currents, were likely to come ashore.

Hereabouts is the likely site of Execution Dock. It was never a dock in the modern British sense, but a grim joke about the place where many pirates ended their voyages. It consisted merely of a gallows on the foreshore. After pirates were hanged, their bodies were left to be submerged by three tides as a warning to all passing mariners. The chain and padlock which secured the condemned men were brought up from the river bed by a PLA dredger in the early 1930s; they were judged authentic by the British Museum.

From time to time PLA dredgers have brought up relics of an older river; bones from prehistoric monsters, ancient swords and other side arms, old muzzle-loading cannon, clumsy medieval anchors, etc. During the last war, one dredger was sunk by an unexploded bomb which went right through the hull. Repaired and back on station, the first object she brought up in a dredging bucket was that same unexploded bomb, which was rendered safe while suspended half way up the bucket ladder.

It was some time in the 1930s that I visited the Turk's Head in the High Street; an unpretentious little Wapping pub left stranded when an older Wapping disappeared beneath riverside warehouses. Built nearly three hundred years ago and closed shortly before the Second World War, it had traditionally supplied a quart of ale to each pirate on his way to execution on the foreshore.

Leading from Wapping High Street to the foreshore are Wapping

Old Stairs. They have for long been too decrepit for safety and are flanked by a later stairway, now also quite venerable. The Old Stairs were worn away by the feet of generations of watermen and seamen, including Nelson himself. At the head of the stairs is a pub, the Town of Ramsgate, formerly the Red Cow, where the infamous Judge Jeffreys was seized by a mob of London apprentices as he was preparing to escape from the country.

At the end of the High Street is Wapping Wall, another ancient way, noted for one of Thames's best-known pubs, the Prospect of Whitby. The name derived from the old Yorkshire-London trade of coasting colliers which, when the Prospect was built some three centuries ago, filled the Pool with a coppice of masts and spars. Between the wars, it was a homely old watermen's pub. Later it had become a show place for tourists.

On one of the two verandas jutting over the river on tide-washed supports, I sometimes used to lunch, while Jenny, the pub's pet monkey, rattled her chain along the rail. Jim Bean, then the landlord, once rolled back a carpet to show me the sockets for ring posts, relics of the prize fights held there when bare-knuckle contests were driven underground.

5

On the Rotherhithe side of this reach of the river, two pubs used to attract me. The Angel, like all riverside taverns, a grand place to watch the panorama of shipping and craft, still boasted the Captains' Room, featured in Walter Besant's novel of that name. Further east was the Spread Eagle and Crown (now renamed the Mayflower), then the only London pub officially permitted to sell postage stamps; a claim that held good until a few years ago when this concession was extended to other taverns. It had originally been granted for the benefit of seamen, Rotherhithe, like Wapping, being for centuries a nursery of the Merchant Navy.

At nearby St Mary's Church, Christopher Jones, master of the Pilgrim Fathers' *Mayflower*, is buried. It was at Rotherhithe, too, that the old ship was broken up. And a little way downstream, still on the south side of the river, is the pedestrian entrance to the Rotherhithe Tunnel, known to dockers as 'the pipe'.

Plate 5:

The Prospect of Whitby, one of Thames-side's best-known pubs. The name derived from the old Yorkshire-London trade of coasting colliers which, when the Prospect was built some three centuries ago, filled the Pool with a coppice of masts and spars. *Photo: Stanley.*

Plate 6:

Towering warehouses of the London Dock which used to be the Port's chief storage centre. In the foreground of the picture is the Gauging Ground where casks of wine paused before being stored in the London Dock vaults. Today, most imported wine is handled in bulk. *Photo: Stanley.*

Back on the north side of the river, Wapping Wall ended at the Shadwell Lock of the London Dock. Beside it is a green pleasance amidst the grey streets of Dockland - the King Edward VII Memorial Park which replaced the old Shadwell Fish Market. It has proved a haven, beneath the plaque commemorating the voyages of Willougby, the two Boroughs and Frobisher, for old sailors of Wapping to sit and weave dreams about the tidal traffic slipping past.

The Lower Pool ends on the north bank with the Regent's Canal Dock, now awaiting redevelopment after a century and a half as gateway to the Thames from the Midlands. It was never placed under the jurisdiction of the PLA since its function was quite different from that of the dock groups taken over by the Authority.

chapter three

Limehouse Reach and Surrey Docks

1

THE LOWER POOL makes a bend into Limehouse Reach where, behind the waterfront on the north side, I found a rich haul of Thames-side interest centring on Narrow Street. Today it is becoming almost a fashionable residential area, but it was then a sordid little way that has nevertheless run through much of England's sea story.

Raleigh walked it to look at his ships; Humphrey Gilbert and William Borough lived in or near it; William Pett built the *Greyhound* there and lived nearby. Duncan Dunbar, of whom more a little later, and Captain James Cook, names which carry more weight in the Antipodes than in present-day Limehouse, lived there. And perhaps typical of this country's prodigality in giving away its technological skills, Admiral Togo worked at a Narrow Street dockyard in 1870 and took what he had learned back to the new shipyards of Japan.

The riverside here matched this background. The old unspoiled watermen's pub in Narrow Street, the Bunch of Grapes, with its riverside stairs, claimed to be the prototype of Dickens's Six Jolly Fellowship Porters in *Our Mutual Friend*. Another claimant was the Prospect of Whitby further upstream. But Number 98, Narrow Street, which was once the Two Brewers, was, according to George F. Young, writing in *The Dickensian* in 1935, the most likely to have inspired Dickens. A pleasant watercolour in the Port of London Authority's collection showed the Two Brewers built out over the river on a crazy-looking wooden structure, but the riverward part of the premises has long since disappeared.

Number 98 had become part of the barge-repairing establishment of Messrs W. N. Sparks and Sons - a long river frontage of rambling buildings and waterworn stone, the longest stretch of commercial

Thames-side surviving almost unchanged from the London of some two centuries ago.

When the firm showed me some bricked-up ovens, I hoped I might have rediscovered the lost birthplace of the famous blue-and-white Limehouse Ware, but it seems they were merely relics of a former sugar bakery, probably founded by sugar bakers from Hanover who, early in the last century, created a thriving German colony in this district.

Five generations of the Sparks family had worked on the river - for more than 150 years. The two directors whom I knew, both with a lifetime of barge work, had memories of billyboys, schooners and other types of craft long since vanished from the Port. And they talked of the bowsprits of square-riggers poking through the bar windows of nearby taverns.

After Sparks had gone, the premises were taken over by the lighterage firm of W. J. Woodward Fisher Ltd. (The late Mrs Woodward Fisher, the well-known Thames-side character who ran the business, must have been unique in the Port's long story of the lighterage trade.) I am told that some modernising of the interiors has been carried out, but externally the line of ancient buildings looks much the same. May heaven preserve it from the developers!

Downstream was Duke's Shore - a flight of watermen's stairs and a patch of foreshore. Here, according to legend, there was once a whipping post and a prison 'cage'. As recently as the early nineteenth century, misdemeanours - including vagrancy and 'sleeping out' - were punished by fines, whipping, the pillory or the stocks. Offenders awaiting corporal punishment were presumably confined in the 'cage' - a local lock-up, where they were under the eye of the watchman. At the head of the stairs the old wooden Watch House, dating from the seventeenth century, still survived.

Below Duke's Shore is a dingy little tidal inlet, Limekiln Creek. On its banks were some seven or eight wharves, one of them the original Limekiln Wharf from which the district takes its name. Of particular interest to me was Dunbar Wharf. Here, about 170 years ago, Duncan Dunbar founded his famous fleet (with never a Thames-built vessel, though local shipbuilding flourished) and the more famous Duncan Dunbar Junior was born in 1804.

The Dunbars fitted out their ships at the wharf and there loaded one of their most profitable cargoes - bottled beer from the Barley Mow Limehouse Brewery for thirsty pioneers in the then new Australian colonies. The bottles had special labels embodying the Dunbar house flag, and it was said that the Dunbars floated to success on Taylor Walker's ale. When I last visited the wharf soon after the end of World War Two, the former rigging loft of the Dunbars still had thick layers of Stockholm tar and rope yarn ends on its floor.

Beyond the Creek, still on the north bank, was the Union Dry Dock. When the first dock here was built in about 1845, the owners solved the problem of providing firm foundations by using the old East Indiaman *Canton* which they sank in position and pinned into the ground. Decks, beams and carlings were removed and her stern was replaced by lock gates, the ship herself becoming the dry dock. This was almost certainly the *Canton* which, in 1797, in company with an unescorted fleet of East Indiamen off the coast of Java, hoisted the colours of a British warship and bluffed a force of French frigates into a retreat. She served as a dry dock until 1898 when she was replaced by a more conventional structure. Among the vessels overhauled there was the *Cutty Sark*. After the last war, the dock was in turn superseded by a modern slipway for a tug and barge works.

Towards the end of Limehouse Reach on this shore are Britannia Wharf and Coconut Stairs, between which the *Great Eastern* was built. At very low water one used to be able to see part of the old slipway where she stuck for three months before Brunel could get her into the water.

2

Turning now to the south bank, near the start of Limehouse Reach was Nelson's Dry Dock, once the headquarters of Bilbe and Company, owners of the Orient Line of Clippers to Australia (forerunners of the Orient Line of steamers now merged with the P & O). In 1854, one of the partners bought a Dutch galiot, the *Reinauw Engelkins*, of only ninety tons. In her he sent from the dock his brother-in-law, a poor clergyman, and his very large family to Melbourne. The ship is believed to have been the smallest commercial passenger-carrying vessel ever to make the 12,000-mile passage; she finished her days as a hulk in New Zealand.

Also on the south side of this reach was the Burning Ground or, as known to rivermen, Condemned Hole. Established early in the eighteenth century by HM Customs for the destruction of seized contraband, it had long since become the headquarters of the Receiver of Wreck for London. It was a ramshackle little wharf, reached from the landward side through an ancient creeper-clad cottage in Odessa Street where the Receiver lived. Here were stored pieces of cargo salvaged from the river or from the sea by inward-bound ships; all officially described as either flotsam, jetsam or lagan.*

3

A little further downstream was the Greenland Entrance Lock of the Surrey Commercial Docks.† The layout of these docks, the ships and cargoes, the dockers - in fact the whole atmosphere - were a complete contrast to the other, more urban docks groups. At the Surrey Docks could be seen the result of piecemeal building by a number of competing developers, followed by forced amalgamations, that were finally incorporated under a single control. The eleven interconnected dock basins (with varying depths) had water areas ranging from some twenty-four acres to as little as two, and names that were pleasantly evocative of the sources of some of London's traditional trades.

Between the world wars, the PLA made many improvements at this group, the most notable being the new Quebec Dock, opened in 1926.

There were three entrance locks to the Surrey group from the Thames: the chief, the Greenland Lock, used by the largest vessels trading here (from North America); the old Surrey Lock, further upstream in the Lower Pool; and the South Lock, below the Greenland Entrance. There was also a connection with the Grand Surrey Canal, whose four miles

---

* *flotsam* - goods that remain floating after a wreck;
  jetsam - goods that have been jettisoned (to lighten a ship in danger) and have sunk;
  lagan - jettisoned goods that have sunk and been buoyed in the hope of ultimate recovery.
† Closed in 1970.

of waterway to Camberwell and Peckham were all that had materialised out of a grandiose scheme of the Canal Era to link the Thames with Portsmouth.

The principal basin, the Greenland Dock (on the site of the old Howland Great Wet Dock, London's first enclosed wet dock), had once been the home of the Thames Arctic whaling trade. After the Second World War, pieces of whalebone found on the site were imaginatively displayed in the outer walls of some of the post-war warehouses at these docks. In the early 1900s, the famous engineer, Sir John Wolfe Barry, of Tower Bridge fame, enlarged the Greenland Dock and its lock.

When I first visited this dock during the late 1920s, it was handling much general cargo, while its refrigerated stores provided a centre for imported dairy produce. However, most of the scattered tangle of Surrey Docks, with their many cuttings and storage ponds, specialised in London's imports of softwood timber and bore such suggestive names as Canada, Russia, Norway, Stave, Lavender and Acorn.

Although deep-sea vessels brought quantities of North American pine and spruce, most of the timber handled there came from the Baltic in a mixed fleet of small ships under charter. It was almost entirely a summer trade when the thaw-out of ice-bound loading ports allowed the ships to sail with perilous-looking deck loads of deals, battens, scantlings, boards and so on. Many of the vessels had an apparently frightening list (due to being loaded while aground at low water) though this did not necessarily cause the dockmaster to refuse them entry except in extreme cases, since their cargoes made them virtually unsinkable.

During the twenties and thirties, this fleet of thin-funnelled, rust-streaked sea gipsies was accompanied by a number of 'onkers'. That was the London River name for sailing vessels in the 'firewood' trade - short ends of deals, battens and boards used in construction work, and not, as the name suggests, for kindling. Many of these onkers were so old and careworn that, despite their buoyant cargo, they were only kept afloat by a windmill pump revolving endlessly on their decks. The pumps squealed and groaned with a noise like 'onk-urr' from which, of course, they took their name.

Some of these ships had been deep-sea fliers in their time and it was pitiful to see them reduced to this last strait of 'sail'. Of the many onkers I noted from time to time in the 'firewood' trade, I can now remember only three barques - *Alastor, Germaine* (ex. *Oakhurst*) and *Shakespeare*;

and I cannot recall whether any of the three were equipped with the complaining pumps.

4

The Surrey Docks men who handled these cargoes were a tightly knit clan, working in even more tightly knit gangs; most of them living close by. Their skills, usually handed on from father to son, consisted mainly of weight lifting and balancing, so that each man had to be something of a tight-rope walker. The lengths of timber were piled to marks and numbers according to the importer's instructions, and as the pile grew - sometimes up to forty or fifty feet - the single-plank staging, about eleven inches wide, up which the men strode ('ran' would be a better word, for they were all on piece-work rates) with their loads, rose with it. Thus a deal porter, as these men were called, would balance a complete fir scaffold pole or a number of planks, perhaps some twenty feet long - the load often not far short of a hundredweight - across his shoulder and the base of his neck.

In spite of the protective leather backing to their hats, a large callous on the neck and shoulder was often the sign of their calling. Synchronising the spring of the staging under his feet and the spring of the load on his shoulder, the deal porter would carry his burden to the growing pile. Small billets would be inserted between each layer for ventilation. Almost literally killing work, it was largely replaced by mechanical handling soon after the war, and, later still, by packaged timber.

Most of the narrow alleyways between the piles of timber were carpeted with sawdust and bark; once the deal porters had moved on to fresh cargoes, the ways were as silent as the forests of origin. Shafts of light slanted across the dim alleys from breaks in the piles and these and the scent of resinous wood invoked a feeling of near-reverence.

Some of the large baulks and logs of imported timber were kept, stapled together into rafts, in storage ponds. Consignments were all clearly identified and the dockers who walked so surefootedly over the dipping and swaying timbers practised many of the skills of lumbermen.

Clean, quiet and fragrant, these docks proved attractive to a variety of wild life, including that lovely migrant butterfly, the Camberwell Beauty or Mourning Cloak (so called because of its dark, crepe-like under-surface).

In the storage ponds and basins were many small fish and several species of waterbirds - wild duck, swans, gannets, moorhens, seagulls and an occasional heron. The ducks, in particular, had an excellent intelligence service that told them which stacks of timber would not be disturbed by deliveries and under which it was therefore safe to breed. (This industrial news service was also noticeable among tideway swans; forty or fifty would gather round Bellamy's Wharf in the Pool ready for gleanings which would drop from a bulk-grain carrier not due to arrive until a later tide.)

5

On this riverside between the Surrey group's old South Dock Entrance Lock and the Greenland Entrance was the Dog and Duck tavern (later 'blitzed', leaving only Dog and Duck Stairs). Its name is said to commemorate a local water 'sport' whereby spaniels chased pinioned wild duck which tried to escape by diving. The tavern could be reached only by a footpath which, it was claimed, had been in use for over three centuries.

Downstream was the Deptford waterfront with its historic creek; a place of splendid memories, though little of the fabric of former glories remained. Professional historians have recorded most of the story: the beginning of the Corporation of Trinity House; Henry VIII's Royal Dockyard (part of which was later fortuitously revealed by bomb damage); its association with Drake and other Elizabethan adventurers; the struggles of Pepys against dockyard theft and fraud; famous ship-builders and their no less famous wooden ships; the explorers who sailed from there to found new colonies.

This was the ground from which, perhaps more than any other, grew the British Empire. Those who now consider this a title of reproach might pause to remember that with it were exported the seeds of parliamentary democracy and the ideals of the British Commonwealth of Nations. And if these, in turn, are tending to become tarnished, the fault lies not in the conception. Rather it springs from human fallibility and reluctance to wait on experience before trying to reshape destiny.

The charming seventeenth-century Trinity Almshouses seem to crouch in the shadow of the unlovely Greenwich Generating Station. *Photo: Stanley.*

chapter four

Greenwich and the India Docks

1

WHERE LIMEHOUSE REACH merges into Greenwich Reach, memories of the vanished glories of Deptford give place to tangible reminders of the Greenwich story, hardly less famous.

The jewel of this waterfront and, indeed, of all London River is the Royal Naval College. Many times have I voyaged round the curve from Limehouse Reach and watched the four blocks of the college sweep into view, two by two, their aloof and spacious dignity mellowed by the graceful arc of the river. Its associations reflect colourful periods of English history. The King Charles Block was built by John Webb, son-in-law and pupil of Inigo Jones, in the reign of Charles II on the site of the former royal Palace of Placentia. Nothing more was done until William and Mary commissioned Christopher Wren to design a hospital for seamen, incorporating the one completed block. In the main, Wren's design was followed, although the work was not finished until the reign of George IV, several years after Wren's death. It remained a seamen's hospital until 1869 when naval pensioners were moved elsewhere, and in 1873 the Royal Naval College, formerly at Portsmouth, moved in. Of distinguished seamen linked with the buildings, the most renowned was Nelson whose body lay in state in the Hospital before burial in St Paul's Cathedral.

My memories of the college, however, stress the by-ways rather than the mainstream of its history; the narrow Tudor bricks in the crypt of the Queen Anne block, survivals of the Palace of Placentia; the two secret doors disguised as bookcases in the King Charles block, perhaps used by pretty ladies making clandestine visits; the self-portrait of Sir James Thornhill on the west wall of the Painted Hall holding out his hand in a gesture which could be interpreted as one of supplication. (He was woefully underpaid for his Baroque masterpiece.)

Nearer our own time is a modest tablet in the stone floor of the entrance to the King William block. Dated 15 June 1941, it records that: 'On this day came three citizens of the United States of America, the first of their countrymen to become sea officers of the Royal Navy.' Could any other service have said it better?

One of my last personal memories of the Painted Hall was in 1962 as one of the guests at a glittering affair of uniforms, decorations and music, the highlight of which was a parade of chefs carrying an enormous baron of beef, escorted round the hall by cadets dressed as seamen of Nelson's day to the music of drum and fifes. Famous ghosts mingled with us as we dined that night.

The Navy has been threatened with eviction. Whoever or whatever the successor, the Thames would suffer an irreparable loss.

If the college presents London River's finest frontage, the background is also worthy of the scene. Viewed from the tideway, it is pure theatre. Behind the college the ground rises to the beautiful Queen's House, the work of Inigo Jones, and the first English house in the modern sense with an entrance in place of the former open courtyard. Now it has been absorbed in the National Maritime Museum.

I saw King George VI and Queen Elizabeth, accompanied by the youthful Princess Elizabeth, carried down river in the Royal Barge to open the museum in 1937. Its models, pictures, instruments, charts, books, medals, uniforms and other relics portray a good cross-section of our national sea story in peace and war.

A relatively new addition to the museum is the old Royal Observatory, Flamsteed House, opened after renovation in 1960. The Observatory had been driven out of Greenwich in 1946 by London's nightly glare and smoke-laden atmosphere to be installed at Herstmonceux.

The panorama of more colourful times flanks the college on the Greenwich waterfront. Where the Ship tavern stood until it was 'blitzed' is now the old clipper *Cutty Sark*, home and dry for good, mainly by the efforts of that devoted lover of 'sail', Frank G. G. Carr, then Director of the National Maritime Museum. The most celebrated survivor of merchantmen under sail, she was built in 1869 as a square-rigged ship for the China tea trade. The *Cutty Sark's* figurehead represented Nannie, the witch in Burns's poem 'Tam O' Shanter', a cutty sark, i.e., a short vest, being all that the uninhibited Nannie wore.

The new ship was dogged by bad luck. First on the China run and then in the Australian wool trade, she made slow passages and earned poor freights. A murder, a suicide, mutiny, cholera and a succession of unsatisfactory masters gave her a bad reputation.

Her luck turned, however, when first Captain F. Moore and then Captain Richard Woodget took over command. During the next ten years she beat all the other clippers, including the famous *Thermopylae*.

In 1895 the ship was sold to Portuguese owners. Later, after serving at Falmouth and then in the Thames as a training ship, she was permanently housed at Greenwich by the *Cutty Sark* Preservation Society. In 1957 she was opened to the public by HM The Queen. To date, more than seven million people have visited her. Among the nautical treasures on board is a number of old merchant ship figureheads. (I shall pick up the origin of this valuable collection when I reach Gravesend.)

Cutty Sark has since been joined by *Gipsy Moth IV*, the yacht in which the late Sir Francis Chichester circumnavigated the world single-handed under sail along the 'Clipper Way'.

The Ship tavern was once the rival of its near neighbour, the bow-windowed Trafalgar (back in business after being left derelict and then, for a time, turned into flats), as the setting for ministerial whitebait dinners of the last century. Lord Palmerston once contemplated a dish of whitebait at the Ship and said to his companions: 'Let us all imitate this very wise little fish – and drink a lot and say nothing.'*

Next to the Trafalgar is the stylish Yacht tavern, very different as rebuilt after being bombed from the pub I knew before the Second World War. Then it was a favourite haunt of Thames watermen.

Further along the Greenwich waterfront, typical of that mixture of picturesque past and utilitarian present so characteristic of the river, are the charming seventeenth-century Trinity Almshouses, seeming to crouch under the wall of the unlovely Greenwich generating station.

---

* Quoted by A. G. Thompson in *Scrap Book of London River*, (Bradley, 1937).

2

From the old Observatory building, the Meridian of Greenwich (0 degrees) runs up Blackwall Reach. Passing into this stretch of river from Greenwich Reach one leaves behind the traditional pageantry and elegant formality of the Royal Navy and finds instead the more homespun activities of the Merchant Service. Blackwall Reach, indeed, has a strong claim to be the birthplace of the British Mercantile Marine. The East Indiamen which berthed here and the Blackwall frigates which followed them were almost more 'Navy' than the Royal Navy in standards of construction, manning and equipment. Not for nothing did the old seamen of the sail say: 'All shipshape and Blackwall fashion.' And after these well-found fleets there followed the incomparable clipper ships which, at their finest flowering, were blown into oblivion by steam.

Pepys knew Blackwall Reach as a place where rich cargoes (and rich pickings) were to be had. The Honourable East India Company knew it as a haven where fortunes, sometimes from a single voyage, were realised. Later merchants knew it as the foundation of most of the City's trade exchanges and commodity markets, chief buttresses of London's commercial power. The West India Docks carried on the story and the way in was via this reach of the tidal river. So, for some three centuries, Blackwall Reach was the main terminal of the best of our merchant shipping.

The West India Docks, opened in 1802, the first enclosed and guarded docks for handling cargo to be built in London, are on the north bank, near the end of Blackwall Reach. When opened, they consisted of two separate docks - the Import and Export Docks - linked by a tidal basin in Blackwall Reach. In 1805, the Corporation of London (then Conservator of the tidal river) constructed a canal across the peninsula formed by Limehouse, Greenwich and Blackwall Reaches, thus creating the Isle of Dogs. But the canal was not a commercial success. In 1829 it was acquired by the West India Dock Company and rebuilt to form another dock, parallel with and to the south of the Export Dock.

The Millwall Dock, an L-shaped basin with an entrance lock in Limehouse Reach (the lock was closed after being 'blitzed') was opened to the south of the older docks in 1868 with an eye to the larger steamships then beginning to supplant sailing vessels. The whole complex of four docks was joined by cuttings and given a new entrance lock in Blackwall Reach by 1929.

The legislation permitting the West India Docks to be built gave the dock company a twenty-one year monopoly of handling all West Indian cargo arriving in the Port, and, although the monopoly had long since expired, this type of cargo still predominated here between the world wars.

As run-down and neglected as the other old dock groups when the PLA took over, the West India Docks had one impressive feature, both decorative and functional. This consisted of nearly three quarters of a mile of Georgian warehouses, originally built as nine separate blocks but soon joined to provide one of the finest industrial façades in this country.

These warehouses were designed by George Gwilt and his son, George Gwilt the Younger, in the face of competition from John Nash, John Soane (of Bank of England fame) and other architects who had already made their contribution to Georgian London. Built of brick, the warehouses were staggered here and there to avoid the monotony of a straight line; there were circular and semi-circular windows, large square loopholes for the delivery of cargo, and ornamental strips of white limestone.

Here were stored the cargoes, sugar in particular, of West India merchants whose near-royal state in London early in the last century was maintained by slave labour on the plantations. These 'nabobs' would drive to the docks in magnificent carriages, behind magnificent horses and attended by a retinue of servants, to inspect the latest cargoes. But whereas the common man had to climb to the upper floors by grimy stone steps, there were broad, mahogany balustraded stairs, masterpieces of the joiner's art, for the nabobs. After the First World War, these stairs, although scratched and stained, still survived.

With true Georgian belief in the immutability of the *status quo*, the West India Dock Company's directors built these warehouses to last for eternity. (In fact, had it not been for their partial destruction by enemy bombs, their massive design might have allowed them to hinder change and so, in a sense, make prisoners of posterity.)

When they were built, in the first years of the nineteenth century, West India cane sugar was imported in hogsheads. With only human muscle to handle cargo on the warehouse floors, these huge casks could be stowed no more than two high. Accordingly, the ceilings were low.

After the First World War, hogsheads gradually gave way to bagged sugar which could, of course, have been stowed more economically had the ceilings permitted higher piling. (After the Second World War, modern machines allowed sugar to be imported loose in bulk and this in turn has now largely supplanted bagged sugar.)

At the West India Docks Rum Quay, more than twenty types of rum were handled; they varied from near-colourless, through pale straw colour to dark brown. Most of this spirit originated in the West Indies, but also came from other countries where sugar cane was grown.

Some thirty thousand oak puncheons of overproof spirit were stowed in vaults through whose deep gloom coopers moved on their frequent inspections. Even the smallest of drips would be regarded much as a major leak in a ship, and the faulty puncheon would be unstowed and repaired. By the 1920s, the coopers were able to use electric hand lamps, but well into the present century they had to make do with polished tin reflectors, catching stray beams of light from the delivery traps (the high inflammability of rum prohibited naked lights).

Before being stowed in the vaults, the puncheons of rum were set out for inspection and Customs' gauging along the quay under a roof which had once provided a covered way to the Great Exhibition in Hyde Park.

On one of the floors were huge vats for clearing and blending the rum. The largest vat held 7,800 gallons; others varied from a few hundred gallons upwards. Like the puncheons, the vats were made of oak and when not in use for rum had to be filled with water to prevent shrinkage by drying out.

Over the Rum Quay and through the vaults hung the penetrating and heady smell of the spirit. But the story of men becoming drunk on its fumes was entirely without foundation.

The other main import at these docks was hardwood with strange-sounding names such as gaboon, jarrah, padauk, sapele as well as the more familiar mahogany, ebony, greenheart and teak. The huge baulks and logs were piled by gantries or travelling steam cranes, and were moved about the quays by horse-drawn trolleys.

The West India Docks had older memories of the glorious days of 'sail'; Joseph Conrad has written about the long line of figureheads and bowsprits overhanging the quays. But, in the 1920s, before the PLA reconstruction programme had got under way, the picture was all grim utility. Unlike the countryfied Surrey Docks, they had scarcely any wild life. But some years later they became the home of an almost legendary fox which shared an undiscovered lair somewhere behind the quays with a homeless mongrel dog. At night or during weekends when there was little going on, the pair could sometimes be seen making the rounds of the garbage bins or carrying off some luckless dock cat.

Outside the dock walls was a tightly packed community of slum dwellers, parochial, inward-looking and dependent for a precarious living on the docks, riverside wharves or ship-repair yards. Shipping had long been the mainstay of the Isle of Dogs, and most of its varied activity was still connected with the sea.

Outside the main West India Dock gates was Limehouse, not the phoney Limehouse which later became fashionable for dinner at one of the respectable Chinese restaurants, but a sordid Chinatown of drug smugglers, prostitutes and gang warfare; where that picturesque pub, Charlie Brown's, was then a meeting place for rough and violent seamen from the seven seas.

In the West India Dock Road leading to the main dock gates was an engineering firm on whose roof was mounted a seven-foot figurehead from some forgotten ship. It consisted of an old-time sailor at the wheel. Chinese seamen at ports around the world would describe Limehouse Causeway as 'the street opposite the sailorman'.

3

In the 1920s, I found the Millwall Dock had much the same shabby presence as the Royal Victoria Dock where I had begun my career. Its main theme was grain, either stored in the PLA Central Granary,* a huge rambling building of conveyors, shutes and silos perpetually shrouded in a fog of grain dust, or delivered to independently owned dockside flour mills.

Square-rigged ships and barques sometimes berthed there at the end of the so-called Australian Grain Race (one of the last refuges of the dying windjammer). During the twenties and thirties, one could occa-

---

* Demolished 1970.

Plate 7:

The Royal Naval College from Island Gardens at North Greenwich. *Photo: Stanley.*

Plate 8:

Some of the nautical relics formerly in the Look-out at Gravesend. The figure-heads are now on display in the *Cutty Sark*; and most of the other exhibits are with the National Maritime Museum. *Photo: Stanley.*

sionally see there the towering masts and spars of such well-known vessels as *Archibald Russell, Herzogin Cecilie, Lingard, Lock Linnhe, Abraham Rydberg* and others which had seen better days. Perhaps these windships, survivors of a nearly extinct beauty, took comfort from the fact that before this dock was built there were no less than seven windmills along the western side of the Isle of Dogs.

When the young men came back to Millwall Dock from the First World War they brought a little colour to this grim area by forming a lunch-time golf club, played over an Emett-like course. Fixed cranes, an abandoned rail truck and other hazards were artfully incorporated, while the holes were the drains at the bottom of warehouse down-pipes. The lunch-time games drew large crowds of dockland spectators, most of whom were convinced that the players were crazy.

4

About half a mile to the east of the West India and Millwall Docks was the little East India Group, built by the Honourable East India Company and opened in 1806 with a monopoly of handling all London-bound cargo from China and the East Indies. It consisted of two docks, the Export Dock, built on the site of the eighteenth-century Brunswick Dock, and the Import Dock, both served by a tidal basin.

In the twenties, the group was heavy with the atmosphere of decay left by the former owners. The march of Empire, which had inspired its construction, was coming to a halt and leaving the East India Docks stranded in modern times. They still harboured ocean-going and coastal vessels and some eastern cargo was still handled there, but the docks were too small and too much out of date to be worth modernising.

However, behind the shabby warehouse walls and scummy water lay a story unsurpassed elsewhere in the Port. Here had sought security from the river pirates first the East Indiamen, then the Blackwall frigates and lastly the clippers; famous and beautiful ships such as the *Crusader, Pericles, Mermerus, Brilliant* and *Lord Warden*. From these later ramshackle basins had sailed many pioneers intent on colonising new lands; mainly family groups who went into the unknown equipped with little more than courage. Sometimes complete village communities, under the leadership of squire or parson, embarked from these quays, and rarely had any of them before seen either the sea or a ship.

The late Captain C. M. Renaut* told me in 1939 about his father's command, the *Crusader*: how bewildered and often lamenting emigrants would go aboard at the East India Docks to the accompaniment of lowing cows and bleating sheep; how the whole scene was one of noise and confusion until his father, in frock coat and top hat, appeared on the poop and restored order. The ship proved lucky and was so much liked that a *Crusader Association* of those who had sailed in her and their descendants was formed in New Zealand.

In the Library of HM Customs I found the sailing lists of the *Lord William Bentinck* and the *Regulus* which left these docks in 1841 for Wellington and Port Phillip respectively. The former ship, only 443 tons, carried 234 emigrants 'equal to 162½ adults', and the latter 369 tons, embarked 'forty three souls equal to thirty and two-thirds adults'. The fractions were the result of calculating the relative space taken up by adults and children, and most of the emigrants had large families.

At the western end of Brunswick Wharf, which then separated the Export Dock from the riverside, was the derelict Brunswick Hotel, built at the beginning of the nineteenth century for passengers and officers of the Honourable East India Company's ships. It was later turned into a hostel for emigrants when its heyday was during the rush to Australia in the 1860s. In the stonework above one of the bow windows the Astronomer Royal of the day had caused the line of the Greenwich Meridian to be cut. (The building was demolished in 1930.)

A record of even earlier emigration was the memorial on East India Docks Pierhead† which commemorated the three little ships that left Blackwall in 1606 for Jamestown, Virginia, where the first permanent English colony in North America was established, fourteen years before the Pilgrim Fathers left this country.

Since the Second World War, the East India Docks have been closed

---

* Captain Renaut was born in Poplar in 1869 and went to sea in 'sail' as Boy at the age of seven. He later transferred to steamships and then left the sea to become Government Superintendent of the Mercantile Marine in Lyttelton. He returned to England in 1939 and died early in the war.

† Later moved to Brunswick Wharf.

by stages.* Near what was the main entrance emerges the Blackwall Tunnel (now duplicated). In the early 1920s it re-echoed more horses' hoofs than car engines.

5

Blackwall Point, at the downstream end of the Reach on the south bank, was one of the places where the bodies of pirates and other criminals were exposed on a gibbet. Here a shelf of rock thrusts all deep-draughted traffic over to the north side.

Between the East and West India Docks on the north bank is Blackwall Yard, claimed to be the oldest continually working shipyard in the world, founded early in the seventeenth century. It reached its peak as a shipbuilding centre during the last century when the firm of Green and Wigram built frigates at and sailed them from the Yard.

In 1939, Arthur Hellyer, then aged eighty-nine, spoke about the days when the Hellyers, the famous family of figurehead carvers (the original figurehead of the *Cutty Sark*, carried away and lost at sea, was their work) had a more or less permanent workshop at Blackwall Yard. He was apprenticed to the craft in 1866 and used to rough out designs in the street beyond the yard. As a boy he had stared with admiration at the gates of the yard, carved, it was said, by the great Grinling Gibbons in the days of Charles II. Unhappily, they have long since disappeared.

The late H. M. Tomlinson, the kindest of friends and the Master Thamesman, wrote lyrically of this area† - 'The chorus of mallets ... in Blackwall Yard': and of Blackwall Stairs:

There one summer evening, as the tide turned again to the sea, I stood to watch a barque cast off. She was bound out. The first lights were moving over the water, green, red and white planets confused in the mirk. The opposite shore had gone. We stared into a void. The barque was spectral as she moved away, and then she dissolved. As we watched the dark where she had been we heard over the water her crew singing a departure chanty at the halliards. Soon, too, their voices went.

---

* Brunswick Power Station stands on the site of the former Export Dock; similar use is to be made of the redundant Import Dock.

† *London River* (Cassell 1951); Quoted by kind permission of Miss Dorothy Tomlinson.

Shipping in the Royal Docks at North Woolwich. On a busy day, up to fifty ocean-going ships might be berthed at the quays. Today, owing to the phenomenal rise of the container trade, dealt with at Tilbury Docks, the Royal Docks are no longer viable. *Photo: Stanley.*

chapter five

Woolwich and the Royal Docks

1

BUGSBY'S REACH (the ugliest name along the river and the source of much far-fetched speculation about its origin) joins Blackwall Reach with Woolwich Reach. Its only interest for me lay in Blackwall Wharf which has a frontage on Bow Creek, estuary of the river Lee, as well as on the Thames. Here was the depot with modern engineering workshops of the Trinity House Corporation. The wharf was often littered with huge navigation buoys under repair, while alongside was sometimes an estuary lightship temporarily off station for overhaul. (Today, the wharf concentrates on the mechanical and electrical maintenance of the Lighthouse Service. The work there adds a piquant flavour of electronics, optics and automation to the gold braid and ceremonial of this very ancient corporation.)

Woolwich Reach in my day was noted for the old steam paddlers of the Woolwich Free Ferry Service (owned by the London County Council, now the Greater London Council). With their tall spindly funnels, looking like Mark Twain's Mississippi steamers, they were magnets for East London boys who sometimes rode backwards and forwards for hours - or until they were chased off by the crews. The attraction was not so much the passing ships as the fascinating steam engines which could be watched from the ferries' alleyways.

The finest hour of these paddlers was during the war when, steaming back and forth all one night, they evacuated to the other side of the river the inhabitants of Silvertown whose isthmus home, between the Royal Docks and the Thames, was threatened by a ring of fire. The diesel ferries that have now replaced the steamers are more time-saving because (with their new terminals) they are designed for the motor age. Yet, inevitably, they lack the nostalgic appeal of the old paddle-wheelers.

The north bank of Woolwich Reach is a heavily industrialised, stinking, unlovely waterfront. But the south shore has royal memories. At the former Royal Woolwich Dockyard, founded by Henry VIII, were built many illustrious ships; Henry's own *Great Harry* and Charles I's *Royal Sovereign* among them. Expeditions fitted out here included that under Captain James Lancaster who carried 'Letters of Recommendations' from Elizabeth I. His successful voyage to the East was made on behalf of the first English East India Company and so laid the foundation of the Honourable East India Company with its trading privileges and administrative responsibilities for British policy in India.

The dockyard was closed in 1869 and became the home of Naval Ordnance. It is now much shrunken owing to the gradual spread of industry. I know of one enterprising company - perhaps there are others - which has installed modern machinery in an old dockyard storehouse that was built by Napoleonic prisoners of war.

Beyond is all that remains of Woolwich Arsenal, now reduced to its core of buildings. On the north side of the river, the Reach ends a little way upstream of the King George V Lock, the main ship entrance to the Royal group of docks.

2

The three interconnected docks of the Royal group (in the Borough of Newham) - the Royal Victoria Dock, scene of my initiation into the PLA way of life, the Royal Albert Dock and the then new King George V Dock, opened in 1921 - were at that time the show piece of the Port. They handled the cream of the country's overseas trade, and their shipping lines served every part of the world, either direct or by transhipment. On a busy day, when up to some fifty ships might be berthed at the quays, there would be a double line over three miles long of ocean-going vessels. The water area of these docks totalled some 250 acres (since slightly reduced by reconstruction of the Royal Victoria Dock).

One of the main features of the Royal Victoria Dock was the Exchange Rail Sidings, then said to be the largest in Europe. Their vast network of permanent way connected the Royal Docks with the country's main lines. Protection of cargo waiting overnight in the rail trucks was a constant problem for the PLA police, since ingenious cargo thieves had even thought up such devices as drilling through the floors of padlocked wooden trucks and on into casks of spirit so as to drain them into strategically placed receptacles.

In some of the tobacco warehouses, huge cathedral-like shells, there was a suggestion of the mills of God in the working of the massive roof gantry cranes which slowly travelled to and fro, stowing the half-ton hogsheads of Virginia leaf tobacco. In other warehouses, tobacco, in leaves of varied shape and size and each with its distinctive aroma and special blending qualities, came from Africa, Greece, Turkey, China, Sumatra, Borneo, Japan, Korea, India, the West Indies and even from Ireland.

Among the cargo handled there from all over the world were scrivelloes (young elephant tusks); spads (young ostrich feathers); cardamoms (seeds used in curry powders); cinnamon; land and sea shells under their different names; a score or more of different gums and resins; and innumerable herbs, leaves, canes, grasses, roots and fibres, as well as skins, preserved foods, essential oils, wines and spirits, frozen and chilled meat.

Cases of manufactured goods from the Far East sometimes proved to be Pandora's boxes by reason of the loose description given by the consignors. Unsuspected strands of dutiable silk in the fabric of dolls' dresses in a case of toys, for example, would cause much trouble with HM Customs when they were not declared. One Japanese importer of exotic foodstuffs in this country would cheerfully endorse his documents with: 'I declare this goods was unsweeting', whether or not he was sure about the dutiable sugar content. When there arrived what I believe was the first case of Mah Jong sets to be imported into Britain, the terse description 'Chinese game' caused it to be entered for Customs' purposes under the heading 'Poultry and Game' and sent to the Cold Store.

3

But the fascination of these dock warehouses lay not only in the strange goods they held, but also in the skilled activities of the staff, which I touched on in my earlier description of the London Dock. It was salutary for us young men (inclined to be somewhat condescending) to watch a sampler, with little education but much experience and a profound interest in his work, draw, say, a sample from a hogshead of leaf tobacco. His selection of good, medium and poor quality leaves would be representative of the complete half-ton of tobacco in the hogshead. Such was the integrity of these men that whole consignments of many hundreds of tons were usually bought on the evidence of their samples alone.

On the quays, alongside the slings of cargo coming ashore, an even older routine was practised – that of the tally clerk. He would put a vertical stroke for every package that passed him, drawing a diagonal line through every file of four; the sign of the gate. It is one of the earliest symbols of commerce, in use long before the most rudimentary form of documentation had evolved, and goes back to the days when man first learnt to count on the fingers and thumb of one hand.

Other primitive and effective methods were adopted, too. I once saw a flock of sheep for export milling about on the quay and refusing to be driven up the ship's loading ramp. The foreman stevedore knew exactly what to do. He picked up the ram and climbed the slope, followed by the whole flock.

One special skill deserved the deepest respect – loading a ship (before the container era). In a vessel bound for, perhaps, half-a-dozen ports, cargo for the first call had obviously to be last in the hold; and cargo for the other ports loaded in order of discharge. But a mixed assortment of cargo would arrive at the dock from all parts of the country almost up to the hour of sailing. Then, again, heavy and awkward packages, say, a case of machinery, could not be stowed on top of fragile crates. Goods susceptible to taint needed to be segregated from pungent neighbours. Lastly, the varied collection of boxes, crates, sacks, barrels, cartons, etc., had to be stowed so that it would not shift in a gale, while still ensuring that the ship remained on an even keel, neither down by the head nor by the stern.

Exports in those days were the poor relation of our overseas trade, at least so far as the Port of London was concerned. Nevertheless, most ships loaded some cargo outward, even if they did not always sail with full holds. 'Striking' (the term for unloading land transport) was done by hand and, although some electric runabout trucks were appearing on the quays, the transfer of export cargo to the shed, ready for stevedores to start loading the ship, was still almost entirely done by hand truck and human muscle.

A humble but valuable part in the export trade was played by the port marker. To help discharge the ship in foreign ports where illiterate labour might well be employed, the port marker daubed appropriate splashes of colour on the packages before loading. Thus all parcels marked with a certain colour would be for, say, Colombo, while another colour would be for Bombay, and so on.

4

On the south side of the Royal Victoria Dock was the Bulk Grain Department. Where there was later to be a continuous line of modern quay, flanked by transit sheds, mills and warehouses, there was then a large area of marshy grass occupied only by flour mills (owned by commercial millers), grain silos and the then new Bulk Grain Office. Its predecessor, together with a granary and storage sheds, had been destroyed in the Silvertown Explosion when a munitions factory, just outside the dock premises, blew up, killing 69 people and injuring some 400, as well as destroying mills and other industrial concerns. The office order book, recovered from the debris, contained a grim reminder. Its page for 19 January 1917 (the date of the disaster) was laced with slivers of glass and stained with ink and blood.

When the dock was built by the Victoria Dock Company in 1855, the marshland had been hopefully included to provide grazing for imported cattle, a trade which never materialised. The mills, with a deepwater frontage, had been added later. Lurid tales were told about what the lusty and uninhibited mill girls had done to men who tried to molest them during the train journey to Silvertown.

Grain discharge by suction elevator had become general but the men who manipulated the elevator pipes were still called corn porters; some of them could remember when grain ships were discharged by basket, shovel and hoist. To the other dockers, these men continued to be known as 'toe-rags', from the strips of cloth with which they kept the grain out of their boots.

To save a walk round half the dock from the north side, we used to ferry ourselves across in the 'sack boat', a superannuated waterman's skiff with a gaping hole in the stern. Embarking at one of the rotten wooden jetties, which in those days projected like decayed teeth from the main quay, we - oarsman and passengers - complete with bowler hats, spats, umbrellas and other status symbols, had to crowd up to the bows to lift the stern out of the water and so keep ourselves afloat and dry.

This marshy area provided a favourite baseball pitch for Japanese crews of the Nippon Yusen Kaisha line. Their ships berthed on the north side, and the teams would ferry themselves across the dock in their own exotic-looking surf boats.

Sometimes we were stirred by the sight of a windship with grain in bulk being towed to the mills. *Beatrice, Hougomont, Lawhill, Magdalene Vinnen* - their names are still evocative of tall masts and standing rigging outlined against the white walls of the mills.

One vessel, the *Grace Harwar*, was claimed to be the last deep-sea square-rigged ship in which, because she was equipped with halliard tackle instead of the more modern halliard winches, the topsail chanty was still sung.

Incidentally, these great sailing vessels inspired me to learn - a seemingly academic but later, in 1939, most rewarding exercise - all about their sails and rigging. Eventually I could tack and wear ship (in theory) with the best of the shellbacks.

In the middle 1930s, the PLA began the complete rebuilding of the Royal Victoria Dock, demolishing the old wooden jetties and other features of my youthful background. The work was sufficiently advanced for the new quays and sheds to play a significant part in loading the D-Day Armada that sailed from London.

5

The Royal Albert Dock extended eastward through the Connaught Cutting from the older Royal Victoria Dock. Its main features were the cold air stores where more than 300,000 carcases of frozen lamb and mutton were normally held. All this meat, as well as chilled beef dealt with at direct-delivery berths, was rigorously examined by the City Corporation's Port Health Authority.

Apart from refrigerated stores, the Royal Albert Dock had no warehouses, being equipped only with transit sheds. The transit shed was an integral part of all dock berths for in it import cargo was sorted to merchants' marks and numbers for delivery, and exports were collected into port consignments for loading. Without the shed, the berth would have been useless to most incoming ships, and undelivered imports or shut-out exports (i.e., left behind in the shed for some reason when the ship sailed) were constant problems for the PLA.

The King George V Dock, linked in parallel through the Adelaide Cutting (now called New Cutting) with the Royal Albert Dock, had transit sheds on the south side and warehouses on the north. A novel feature of these warehouses, the underslung roof cranes, could carry out five different operations - hoisting, slewing, traversing, travelling and derricking. Although much work at the King George V Dock was still manual, these cranes hinted at the pattern of things to come.

6

It was boasted that the Royal group of deep-water basins was only five miles from the heart of the City; which, as transport by water is cheaper than by land, was counted a considerable advantage to London merchants. What none of the planners foresaw was the upsurge of motorised road traffic that was to choke the approach roads and so render this economic advantage an illusion. The Silvertown Way, opened in 1934, cut through an area of appalling slums and provided a new motor route to these docks. But almost from the start, its effect was largely offset by the congested traffic arteries leading to it.

All road entrances to these and other PLA docks were guarded, day and night, by dock police, and every scrap of cargo leaving had to be accompanied by an official pass.

The water entrances - the locks - by which ships and craft entered and left the basins, were then four in number - the new King George V Lock, Gallions Lower and Upper Locks leading to the Royal Albert Dock, and the Western Entrance to the Royal Victoria Dock. They were all the special charge of the dockmaster and his assistants. He was also responsible for co-ordinating all movement afloat in the vast lagoon provided by the three Royal Docks; for the passage of ships into (called 'stemming') and out of the three dry docks; and for the swinging or lifting of all moveable road bridges, with due regard to the crossing rights of pedestrians and road traffic. And the crux of all this work - the vital sea gates of these docks – were the four locks. Through them passed each year nearly three thousand ships and more than thirty thousand barges.

The Royal Docks are impounded by pumps to a level of two and a half feet above the average high water in the river outside. As the Thames here rises and falls some twenty feet, locks are virtually gigantic lifts to raise ships up to the level of the dock water or lower them down to the river.

The main water entrance to the Royal Docks is the King George V Lock. Lacking a ship for comparison, it could look rather insignificant – somewhat like a largish swimming pool, although, in fact, 800 feet long and 100 feet wide. But the entry of a ship would put the lock into true perspective.

When the rising tide had given sufficient depth of water over the lock cill, the blue docking flag would be hoisted to signify that it was ready for customers (all previously booked by telephone in the dockmaster's lockside office).

Two tugs would deftly position the ship in the river off the lock and hold her there against the tide while a waterman's skiff ferried a headrope ashore. From then on, the dockmaster was in sole charge, conducting all manoeuvres by means of a whistle, similar to that of a football referee, and hand signals and never a shouted order.

With engines going slow ahead, the ship would enter the lock, one of the tugs hanging on to her tail to hold her against the sweep of tide across the entrance and not casting off until the ship was in the slack water under the pierhead. To the uninitiated, it would seem like an attempt to put the proverbial quart into a pint pot, but rarely would a ship even graze the lockside.

The dockmaster would walk backwards watching the incoming ship, his whistle continuing to control the vessel's engines and his own locking crew who marched beside the slowly moving ship carrying huge check ropes like zoo attendants with a python. These ropes would be whipped on and off hydraulic capstans or bollards as directed by the whistle.

A series of chirrups by the dockmaster would indicate that the ship was in position and could now be made fast. More whistles, and the gates would be shut and the sluices opened to allow the lock to fill to the level of the dock water far above the ship.

The ship's bridge and boat deck, only a few feet above the lock side, would begin to rise slowly until, some thirty minutes later, the vessel would be towering over the lock walls. At this point the Trinity House (or 'mud') pilot would disembark and the dock pilot would take his place. The writ of Trinity House does not run inside PLA docks, the authority of the river pilot ending or beginning only at the lock. The dock pilot is therefore merely an experienced guide (although invariably a licensed waterman) privately employed by individual shipping companies.

The whistle would at last bring attention back to the ship in the lock where the water would now be level with that of the dock. Traffic over the road bridge spanning the lock would be halted (frustration of meeting a 'bridger' was a local traffic hazard) and a further whistle would cause the bridge to divide into two arms rising to clear a way for the ship. The inner lock gates would open and two PLA tugs would back in and make fast to the vessel. A last whistle, and the ship would move slowly out of the lock to take up her allotted berth in one of the three docks. It would all be done as smoothly as a hand sliding into a silk glove.

Each PLA dock group has its own dockmaster and his assistants, certificated deep-sea master mariners who, like the Trinity House pilots, have abandoned foreign-going seafaring. And, like the pilots, their hearts are often rent when they see a fine ship outward bound.

7

I was at the Royal Docks in 1926 when, on 3 May, the General Strike began. The Trades Union Congress called a general strike in support of the Miners' Federation which was resisting an attempt by the coal owners (then private individuals and companies) to impose lower wages. Some two million workers came out. So far as London was concerned, the focal point was these docks where huge quantities of imported meat and other essentials were stored.

A destroyer moved in; a detachment of the Brigade of Guards, complete with machine guns, took station at the dock gates; some hundreds of volunteer labourers lived aboard passenger ships moored at dock quays; and armed convoys of food lorries shuttled between the flour mills and warehouses, in particular the cold stores at the Royal Albert Dock, and the main food depot in Hyde Park. When the electricity supply to the Royal Docks was cut off by the socialist-controlled West Ham Council, the cold stores and other plant needing power were kept going by the generators of submarines berthed alongside.

Groups of sullen strikers glowered at us as we entered or left the docks but, personally, I saw no violence.

8

Immediately beyond the dock gates were some of the worst slums of dockland where, during the late 1920s and the 1930s, I saw a silent revolution on the move. The emancipation of women engendered by the First World War, the growing influence of radio (we usually called it wireless then), cinema and popular press - all these were combining to eliminate the last of the dockside slatterns. Such women, typically wearing their husbands' cast-off caps, secured with hatpins as formidable as themselves, had still been conspicuous when I first arrived at the docks. On warm evenings, they often sat on a chair or a stool on the pavement outside their dreadful little dwellings. Now they were acquiring self-respect and beginning to take pride in their appearance. Today, their granddaughters are among the most fashion-conscious girls in London.

chapter six

Gallions Reach to Tilbury Docks

1

BEYOND WOOLWICH REACH is Gallions Reach where the river widens and begins to hint strongly at the still distant sea. Between the wars, the south bank and beyond was a succession of lonely marshes with few houses and little industry to break the skyline. The scene is changing fast, since the gun-proving grounds of the Arsenal were taken over for the new town of Thamesmead, due to be completed in the 1980s and already a distant contour of towers.

On the north bank, downstream of the entrance locks of the Royal Docks, Beckton Gasworks, with cranes, jetties, slag heaps and retorts, continued the utilitarian atmosphere of the reaches upstream. In the thirties, they were claimed to be the largest gasworks in the world.* Today the manufacture of gas from coal steadily declines and new technology is producing riverside changes.

I have always found difficulty in reconciling this grim shoreline with the large salmon which my father saw killed with a boathook in a low-water pool at Beckton near the end of the last century. More poignant is the memory that off Beckton in 1878 the paddle steamer *Princess Alice* sank with the loss of some 650 lives after colliding with the collier *Bywell Castle*.

Downstream of Beckton in Barking Reach is Barking Creek, estuary of the River Roding. Alongside the Creek is the Northern Outfall of the Thames Water Authority's (then LCC) sewage works. Its counterpart is the Southern Outfall at Crossness on the opposite bank of the

---

* Now closed and awaiting redevelopment.

Thames in Halfway Reach. At both outfalls I saw (I must admit without much enthusiasm) detritus channels, screens, sedimentation tanks, aeration tanks and other unromantic aspects of the scientific treatment of sewage by stages aimed at pouring a comparatively pure effluent into the tideway, leaving the solid matter to be dumped at sea.

Dagenham, on the north bank of Halfway Reach, was for many years noted for Dagenham Breach. In 1707 the Thames had burst through here to flood a large area of marsh. When the river wall was repaired, a big lake remained. Pitt used to fish there, a recreation which led to the ministerial whitebait dinners at a local tavern. The dinners were later transferred to the Brunswick Hotel and, later still, to Greenwich.

In the 1930s, the Dagenham shore was dominated by the then comparatively new Ford factory which gave me my first experience of perpetual motion along the assembly lines. The monotony of some of the work filled me with horror, but my guide assured me that the workers were content to think of their personal affairs while they pulled the handles or levers up or down all day. The mass-production concept, then still something of a novelty in Britain has, alas, become a commonplace and a major source of industrial dispute.

The next focal point of my memories is on the south side of Erith Rands, a short stretch of river linking Erith Reach and Long Reach; where Dartford Creek, estuary of the River Darenth, joins the Thames; where the first of the Measured Mile Beacons stands; where Long Reach Tavern was formerly a landmark.

The Darenth was of old a beautiful little trout stream which, like many other rivers in the south-east, became so badly polluted that for years nothing lived in its waters. Soon after the Second World War, a PLA Harbour Master, like myself a committed fly fisherman, discovered that the Darenth was again in good health. And we planned to rent a stretch and restock it with trout. Our plans fell through, but later on the idea was taken up by other enthusiasts. Now the river is once more Spenser's 'still Darent, in whose waters cleane ten thousand fishes play'.*

The Measured Mile Beacon is a reminder of the days when Thornycroft built small warships for the Navy at an upriver yard. The ships

---

* See page 146 *et seq.*

were stripped down so that they could pass under Thames bridges and be brought to Long Reach for Admiralty speed trials.

My most fruitful sortie into this area was a visit to Long Reach Tavern, solitary in a wilderness of marsh, the loneliest pub on Thames-side. The licencee pointed out an adjoining meadow, still called the 'Fighting Marsh', where bare-knuckle prize fights were staged more than a century ago. When these contests became illegal, they were planned with greater care; sentries would be posted on the river wall, plank bridges over strategic dykes would be taken up, and boats provided ready to whisk fighters and spectators over to the Essex shore should the police appear.

One of these contests lasted the best part of two days and stopped only because of a police raid; one of the fighters ran into the tavern to hide - and dropped dead. Long Reach Tavern was destroyed by fire in the 1950s.

Near the end of Long Reach, the Dartford-Purfleet Tunnel, opened in 1963, links Essex and Kent. It is a strategic crossing of great value to the trade of Tilbury Docks further downstream. During its first thirteen years of operation 106 million vehicles here passed under the Thames, not far from the site of the medieval ferry by which parties of pilgrims to and from Canterbury crossed. The tunnel has been so successful that a duplicate is nearing completion.

Where St Clements Reach joins Long Reach, on the south bank, is Greenhithe, a pleasant symphony of white chalk and green foliage. Off Greenhithe was the training ship HMS *Worcester*, owned (up to 1968) by the Thames Nautical Training College* which also had a shore base at nearby Ingress Abbey. The old ship gave her cadets a sound education as a background to training for the Merchant Navy. It was once a tradition that the last cadet into his hammock at night should be assisted by a rope's end, so that while they were saying the compulsory prayers there was much surreptitious undoing of buttons. The final amen was accompanied by an almost instantaneous shedding of clothes.

Greenhithe is still the home of F. T. Everard's fleet. Its comparatively small vessels, all with names ending in 'ity', constituted an impressive

---

* Now the new Merchant Navy College administered by the Inner London Education Authority. The *Worcester* was sold for breaking up in 1978.

proportion of our coastal shipping. The firm, in fact, had difficulty in finding enough names with the suffix 'ity' which led to the use of some rather unfamiliar nouns at least where ships are concerned, such as Angularity, Suavity, and Aseity. Today, the fleet includes a good number of tankers.

Before the Second World War Everard coasters spearheaded the British challenge to intruding Dutch schuyts (not to be confused with the older type of schuyt selling eels at the buoy off Billingsgate). In the absence of protective legislation, these little foreigners were acquiring almost a monopoly of the carrying trade between British ports. The Dutch schuyt was often the permanent home as well as the livelihood of her crew and it was not unusual to see one, her simple oil engine thudding 'kertonker, kertonker', with broad-beamed mother at the wheel, father and sons busy about the deck, and a line of washing drying on a halliard.

As a postscript to this area, I recall the oldest known Thames-sider who lived a quarter of a million years ago in what is now Swanscombe. Bone fragments found in the middle gravel of the Barnfield Pit fitted perfectly to provide part of what was claimed to be the oldest known human skull discovered in Europe, as indicated by carbon dating. Swanscombe Man was a prototype for *Homo sapiens*, a hunter who made flint axes. Finding the skull was a moment of truth for any present-day man of the Thames with a little imagination; for Swanscombe Man's river, now narrower, possibly deeper and peopled by less horrific fauna, but still the Thames, flows on.

2

Of the Thames's five docks groups, I came to know Tilbury last. For one reason, it was twenty-five miles downstream of London Bridge, and in those days I was discouraged by the few and thoroughly sordid steam trains from Fenchurch Street Station. For another, Tilbury's apparent remoteness from the City of London suggested that the writ of the PLA ran more slowly there than further upstream.

Then Tilbury consisted of a main dock with three branch docks - East, Centre and West - set at right angles; the total deep-water area was about eighty-six acres. The old entrance lock, approached through an open tidal basin from Gravesend Reach had been designed only for the largest vessels afloat at the time of building in the 1880s.

The misguided hopes which had caused the East and West India Docks Company to carve this terminal out of lonely Essex marshland years ahead of its time had been one of the principal reasons for the ruinous competition that led ultimately to the creation of the PLA. The aim was to attract shipowners to a new dock opposite the shipping centre of Gravesend and so much nearer the sea. The London merchants, however, were outraged at the expense of such a comparatively long journey for their cargoes.

Tilbury Docks had then twenty-four transit sheds but no warehouses for storage. With poor road approaches, they relied on barge or rail links for the distribution of cargo, which went either direct to factory or market or to the warehouses of the upper docks; an arrangement that created yet another problem for the dock company. This came from the lighterage trade which claimed that traditional Thames craft were not designed for the more exposed waters of the river at Tilbury and so demanded higher rates in compensation.

The PLA at first considered its inheritance at Tilbury to be too far from the City of London to have much of a commercial future. But the continued growth in the size of ships after the First World War indicated its potential advantages which could be exploited only by adventurous development.

Although money was tight, pressure from the Government, anxious to reduce post-First-World-War unemployment, encouraged the Authority to proceed. On Tilbury riverside, outside the docks, a new landmark, the deepwater, double-decked, rail-connected cargo jetty was ready in 1921, intended mainly for ships calling with part cargoes for London.

By 1927 dredging of the new ship channel from the Port's seaward limit up to London Bridge had been completed. It helped to attract to Tilbury Docks the deeper-draughted vessels coming into service.

A new Tilbury entrance lock (opening off the Northfleet Hope just before it merges into Gravesend Reach), 1,000 feet long and 110 feet wide, and a new dry dock, both still the largest in the Port, were opened in 1929 and have since proved a solid basis for the more revolutionary expansion of the 1960s. The dry dock has unique features in the form of automatic bilge blocks and a leading-in girder which ensures the vessel coming to rest in a central position.

When the dockmaster took me to look at the dry dock soon after it had been opened, it was being pumped out. After the water had receded, he broke off his talk about its mechanical wonders to goggle at a dozen or so large pollack flapping about in the remaining puddles.

Downstream of the cargo jetty on the riverside was a new rail-connected passenger landing stage - a floating structure, 1,142 feet long, where the largest liners using the port could lie afloat at all stages of the tide. It was opened in 1930 by J. Ramsay MacDonald, then Prime Minister, to decorous applause from hundreds of VIPs.

One ship to be seen regularly at the stage before the war was the star of London River, the twin-funnelled *Viceroy of India*, built at the end of the 1920s. Wearing the traditional P & O colours of black hull and buff upperworks, she had beautiful lines with superb passenger accommodation. Notably, she pioneered for her company the turbo-electric drive and lowered the record from London to Bombay on her maiden voyage. More than once I have heard old P & O hands talk nostalgically of her as the best ship the Line ever had.*

People below the age of, say, thirty, who have grown up with a network of world transport whereby one can reach the ends of the earth in a flight of a few hours, may find it hard to appreciate the appeal which the landing stage had for my generation of Thamesmen. Here were the magical promises of far voyaging, the hints of new frontiers, hotter suns and bluer seas. With it all mingled the sadness of separation. When a great ship at the stage gave a compelling blast and the tugs took up the slack, the atmosphere was thick with urgent emotion.

But it was also during the thirties that the owning of a Merchant Marine became virtually a status symbol for any nation with some sort of coastline, and subsidised foreign competition hit British shipping hard. Then the landing stage became the centre for cruising, notably for the £1 a day tourist cruises whereby liner companies sought to avoid laying up their passenger ships.

Tilbury landing stage is also the Essex terminal for the short ferry (so

---

* Lost during the Second World War.

named to distinguish it from the long ferry which, by tide, oars and sail, ran between Gravesend and London from before the Norman Conquest until well into the steamship era). The short ferry between Tilbury and Gravesend has a history nearly as ancient as that of the long ferry, and behind it is a confused story of wrangles over manorial rights and privileges. Both Tilbury and Gravesend are ancient towns, and the short ferry, like the long ferry, has been in existence since at least the Norman Conquest. In 1694, it was taken over from the respective Lords of the Manor by Gravesend Corporation. Later acquired by the London, Midland and Scottish Railway, it is now operated by British Rail.

In more recent times there were two short ferries running between the landing stage and Gravesend. One, for pedestrians, still operates, but with new ships and modernised approaches. The other, for vehicles, was put out of business by the Dartford-Purfleet road tunnel.

Near the landing stage was the solidly comfortable Victorian Tilbury Hotel, built by the old dock company and bequeathed to the PLA. It was designed for a more leisurely era when communications were such that passengers and their friends often stopped overnight at Tilbury so as to be ready for an early-morning embarkation or arrival. A landmark to all rivermen, it was something of a white elephant to the Authority. Nevertheless, it had a small regular clientele who were attracted by the unrivalled panorama of Gravesend Reach, which they could watch through the telescope mounted in the lounge, and also by a very fine cellar. (The hotel was built of timber and made a sadly spectacular bonfire in 1944 when it was completely destroyed by incendiary bombs.)

3

Another and quite different aspect of Tilbury Docks during the 1920s was the extent and variety of their wild life, not surpassed even by the Surrey Commercial Docks. Fired by what had been at the time premature optimism, the dock company had left the Tilbury basins partly surrounded by hundreds of acres of land for future development. This area (today the site of container berths) was clothed in summer with a vast jungle of cow parsley and was something of an unofficial nature reserve. Here and there were strange and sometimes exotic plants which I was not botanist enough to name but which had probably grown from seeds brought ashore with cargo.

The wild life included an abundance of adders that sometimes nested under goods in the transit sheds. Amongst the cow parsley were rabbits, stoats, an occasional fox, partridges, pheasants, hedgehogs and sparrow hawks. Swans, duck and several species of waterfowl used the docks as swimming pools. Farouche cats, adaptable enough to respond to the noon whistle, would emerge from their jungle hide-outs for titbits at the offices.

Asian seamen sometimes varied their shipboard diet with bundles of a sort of wild spinach which they found growing about the docks, and by shrimping round the edges of the basins.

During excavations for later dock development at Tilbury, a stone tablet with a Greek inscription commemorating the wife of a Greek citizen of Roman Londinium was found. The tablet had probably been carried down river in one of the loads of soil removed while constructing London's Underground Railway and dumped on the Essex marshes before these docks were built.

4

East of the docks at Tilbury is a lonely hostelry of character – The World's End. The marshland here is associated with Elizabeth I's famous review of her troops in August 1588. By then the Spanish Armada had fled northwards but there was still the threat of invasion from the Spanish Netherlands, against which the English troops had been assembled.

The World's End was built about the end of the seventeenth century by the officer commanding Tilbury Fort. It seems that his pay was in arrears* and that he received tolls from the Essex side of the short ferry as compensation. No doubt the old inn sheltered smugglers, couriers and spies who crossed the river here as the quickest route to the Dover Road and an unobtrusive way out of the country.

---

* Recorded by the late Frank C. Bowen, the Gravesend historian and author.

The Tilbury Fort, of which that officer was in command and which survives on the riverside, was not the one associated with Elizabeth I. The old fort had been rebuilt, and the beautiful gateway added, during the reign of Charles II under threat of Dutch invasion. Part of the natural defences of the Fort was Bill Meroy Creek. Many unsuccessful attempts have been made to pin down Bill Meroy; yet he remains, like Bugsby, an historical enigma.

chapter seven

Some Ships, Craft and Men

1

BETWEEN THE WARS, many fine ships regularly turned round in the Port of London. They were, in the context of the times, marvels of sophistication, both as to design and methods of construction evolved in the forced draught of the First World War. And the burden of some four decades descends on me when I realise that practically all are gone; some lost during the Second World War and others scrapped as obsolete. Even the best of them would have looked old fashioned beside the vessels of today. Yet then, we thought them perfection.

Dipping at random into my memories, I recall the handsome little *Ausonia*-class Cunarders running out of Surrey Commercial Docks; the fast but plain-looking ships of the American Merchant Line (whose steam whistles were the loudest in the Port), then berthing in the Royal Albert Dock; the sturdy *Beavers* of the Canadian Pacific, with their stout goal-post derricks, also using the Surrey group; the *Britannic* and *Georgic* of the White Star Line in the King George V Dock; the towering superstructures and squat funnels of the Nelson Line's *Highlands* and the *Marus* of Nippon Yusen Kaisha in the Royal Victoria Dock; and the white-hulled *Straths* (*Strathnaver*, *Strathaird*, etc.) with other P & O giants (by the standard of their day) at Tilbury. In the East India Docks one could often see the graceful clipper bows and figureheads of the Ben Line ships. Here, too, was the permanent home of Scott's *Discovery* until she was moved upriver to her Embankment berth.

Looking back to the late 1920s and early 1930s, I am astonished to recall how many ships were still being built with two and even three funnels. The P & O *Strathnaver* and *Strathaird* began life with three funnels apiece (although afterwards reduced to one). The *Georgic*, *Britannic* and the *Highlands* all had two funnels, and there were many others so equipped. But often these supernumerary funnels in the new ships were dummies, used as boatswain's stores. The Orient liner, *Orion*, which came into service in 1934, followed the new fashion of one funnel and one mast.

Steam was still regarded with warmest affection by ships' engineers, but the internal combustion engine was beginning to provide serious competition. The pioneer diesel ship, *Selandia*, had visited the Port in 1912 when she had been inspected by the youthful Winston Churchill, then First Lord. There are no more conservative animals than men of the sea and it took the stresses of the First World War to force grudging acknowledgement that the new-fangled motors were not so unreliable after all. The new *Highlands* (*Highland Brigade, Highland Chieftain*, etc.) were so driven, and the *Georgic* and *Britannic*, at the time of their building, were acclaimed as the world's largest motor ships.

Of the smaller vessels, I recall most clearly the colliers, and especially the 'flatirons' - large sea-going ships with collapsible masts and funnels and vestigial superstructures to enable them to pass under London's bridges in the service of upriver gas works and power stations. The skill of their pilots lay in gauging when the depth of water under a bridge and the air draught between water surface and the soffit of the bridge arch were right for the ship's passage. Usually a margin of only a few inches was available on both counts. In those days the flow of coal from the north-east and south Wales coasts constituted the Port of London's largest import - over thirteen million tons a year.*

There were several coasting liners at that time, and it was possible to book a passage right round the British Isles. One of the most distinctive fleets of coasting freighters was owned by the Carron Line. Their ships incorporated a cannon ball near the top of the main mast, a memento of the days when this fine old firm's Scottish foundry cast the naval gun, the carronade, which helped our wooden walls to rule the seas.

Into the Tilbury Dock Basin ran the little 'rabbit boats' of the John Cockerill Line. They took their nickname from the large quantities of Ostend rabbits and hares which they brought to this country. Up to Billingsgate Market steamed the little fish carriers, a sort of marine Carter

---

* See page 152 for today's comparable figure.

Patterson which collected fish direct from fishing fleets in the North Sea; today the fish comes overland.

Downstream, during the summer, went the 'Butterfly' paddlers of the General Steam Navigation Company (which claimed to be the oldest ocean-going steamship company in the world) taking Cockney London to the sea, often the East Ender's one-day annual holiday.

A sight that had become rare, even in my early days in the Port, was the arrival of a steamer with a band of purple painted along both sides of her hull as a traditional sign of mourning for the death of her owner. This was a custom much followed in the days of sail when owners were more personally involved with their ships.

A part of the river scene in those days was a fleet of small paddle-wheel ambulance steamers which lay at the Metropolitan Asylums Board Pier at Deptford. They were designed to carry patients suffering from smallpox and other contagious or infectious diseases to an isolation hospital near Gravesend. They reflected our forefathers' preoccupation with epidemics and were not superseded until after modern prophylactics and reliable road ambulances were in general use elsewhere.

Another type of craft which has virtually disappeared is the coaling elevator, a floating latticed tower whose grab running along an arm, backwards and forwards between coal barge and ship's bunker shute, caused it to rock drunkenly to and fro.

2

My early experience of the tideway at close quarters came through friendship with the tug owners W. H. J. Alexander, Ltd, whose famous fleet of *Suns* has for long been almost as much a part of the river scene as the tides. Whenever I had an opportunity I would board one of their craft as passenger.

These vessels, unlike the smaller 'toshers' or craft tugs, do not tow for long distances in the river (unless the tow's engines have broken down); their main duty is to help big ships to manoeuvre. They were then all steam tugs and I marvelled at the *rapprochement* between each tug skipper and his engineer. Nudging the tug's 'bow pudding' against the hide of a ship to urge her into her berth, taking the strain of the great rope to help an anchored vessel swing to the tide, giving the tow a sheer to one side - whatever the operation, the 'chief', out of sight in his compact little power house down below, seemed to sense what the next order would be and had his engines obeying the telegraph almost before the gong had clanged. All the same, each engineer complained that his skipper was too fond of 'backing and filling' (the old sailing ship term for going ahead and astern).

Sober looking and rather ugly in those days, with thin upright funnels, these tugs had very broad beams and sterns like the butt end of a toad. But they were uncommonly hard working and efficient; and no one who knew their worth or who had watched them in action would ever use the cliché 'fussy' for such craft. That father figure amongst our scientists, Thomas H. Huxley, once said he thought he would like to have been a tug if he had not been a man.*

The tug skippers, all licensed Thames watermen, were inclined to be taciturn, often paying the price of irregular mealtimes with bouts of dyspepsia. Already they were discarding their bowler hats for the less formal soft felts; and, while a few kept a white-topped uniform cap in their cabin, I never saw one worn in these craft. Some of the engineers had a Board of Trade Certificate but most of them were 'shovel engineers' - that is, they had graduated from the stokehold.

As we rode the tides, nosed into the enclosed docks or waited at the Gravesend Buoys for an expected ship, the crews yarned to me about the good old days when the hazards of 'sail' and the fire danger from old-fashioned oil lamps brought them many lucrative salvage jobs.

These windfalls were known up and down the river as 'hovels', deriving, the crews believed, from the former hovels at Deal. Awakened at night with news of a profitable wreck on the Goodwin Sands, the Deal 'hovellers' would tear through the streets, dressing as they ran, while their womenfolk, up to their necks in the sea, were already launching the boats. My own view is that the term 'hovel' comes from the huffler, the local pilot who used to board sailing craft to help the crew work through the arches of a difficult bridge.

In all these stories the river grapevine was prominent in bringing news of ships in trouble, perhaps twenty miles away, within minutes. But their nostalgic yarns of tugs rushing off with dowsed or even reversed lights

---

* *The Huxleys*, by R. W. Clark (Heinemann) 1968.

to throw rival crews off the track of the hovel, of tugs being baulked of their prey by lack of bunkers or of crews becoming comparatively rich from a single tide's work, were eclipsed in 1930 by the feat of my friend, the late Charles Alexander, then in command of *Sun IV*.

Helped by *Sun XII*, he towed an abandoned tanker, with her oil fuel on fire, head to wind all night to prevent the flames spreading to her cargo of 600 tons of petroleum. During this operation it was learned from the tanker's mate, now a passenger in the tug, that the door of the pump room had been left open and that if this were not closed the fire would eventually detonate the cargo like a bomb. Charles Alexander boarded the tanker and, feeling his way through the flame and smoke, managed to shut the door. Next morning the tug went alongside again and the crew was able to put out the fire.

The Admiralty Court made a substantial salvage award to Alexander and his two crews. When counsel asked him what he would do with the money, Charles's response was characteristic. 'Are you married?' he inquired. 'Yes,' replied counsel. 'Then don't ask bloody silly questions,' said Charles.

3

Returning to my own tug voyages, I liked best the night passages when the harsh and ugly Thames-side chimneys became Whistler's 'campanili' and warehouses 'palaces in the night'. Sometimes a distant red mouth suddenly opened in the darkness of the river bank and then we might see the night shift silhouetted against the furnace's glare like starlings in the setting sun. A retort at some invisible gas works might fling a great sheet of crimson and purple light across the water, perhaps briefly illuminating the sombre figure of a lighterman navigating his cumbersome barge through the crowded fairway. And there was a touch of Edgar Allan Poe in the shadow of a giant piston which lunged to and fro across the lighted window of a factory.

Night or day meant little to the river; it was the tide that ruled. As we wound round the reaches, constellations of red, green and white lights swam with us, our skipper telling me that the invisible ship or craft was the such-and-such piloted by old so-and-so; that she was the Dutchman bound for Custom House Quay or the Cunarder for Surrey Docks. Steam whistles - one, two or three blasts - indicated to invisible

pilots what invisible ships intended to do. And while the great stream
of traffic handled by hundreds of men moved up or down river, my
other world ashore slept, and many of those asleep had little or no idea
that men's lives were not necessarily ruled by the sun.

4

I had mixed feelings about the hundreds of red-sailed barges
which were then so distinctive a feature of London River. I admired the
skill of the two men who handled these deep-laden boxes, with their
huge spritsail, lee-boards (forerunners of the sailing dinghy's centreboard)
and mizzen cunningly sheeted to the rudder to aid the helmsman.

I was also receptive to the philosophy of the tugmen for whom these
beautiful anachronisms were merely a damned nuisance. A fleet of them
tacking across the river in line ahead and prepared to maintain the
principle that sail takes precedence over steam could be a source of
inspired language to the tugmaster making his way downstream, perhaps
with an unwieldy tow 'flying light', that is, unladen and trying to sheer
all over the fairway.

But artists and enthusiasts of 'sail', to whom these craft presented no
navigational difficulties, found them of unique appeal. Most were of
wood, although some steel coasting barges had been built. The average
barge was about eighty feet long with a beam of some eighteen feet and
an incredibly shallow draught - 'sail 'em in a heavy dew' claimed the
enthusiasts. Almost the truth, for they were found up creeks which,
even at high water, could have been waded.

To the uninitiated they were all just barges; to the rivermen they
belonged to well defined types. There were stumpies, without a topsail
and used largely in the brick trade; stackies, hay barges, their loads piled
more than ten feet above the deck, the mate perched on top of the stack
acting as look-out for the skipper at the wheel; and a few boomies with
boom-rigged mainsails instead of the more usual spritsails.

A Rochester barge usually brought a loud 'baaa' from the tug's funny
man, the story going that a Rochester barge hand had been hanged for
sheep stealing. But this legendary insult was likely to be offered to any
passing craft with rural connotations.

Apart from special fleets such as the once-famous Blue Circle Cement
barges, these sailing craft were essentially inshore tramps. Their owners

sought such freights as flour, grain, coal, timber, explosives, etc., where opportunity offered. The crew of two were usually on shares and, when there were no freights, had to make do with 'subs' from the owner, to be paid back when they were earning again. Barges temporarily out of work often tied up at Woolwich Buoys, known for this reason along the river as 'Starvation Buoys'.

Tugmasters swore that if they went within half a mile of a moored sailing barge there would be a wash claim, in which a length of frayed old rope, alleged to have been severed by the swell of the tug's passing, would be produced as evidence.

Like the tug crews, the sailormen (as these bargemen were called) seemed to be aggressively civilian in their dress. I once caught a fleeting glimpse of such a skipper wearing a frock coat, green with age and salt water, and a cap with ear flaps.

5

Sworn enemies of the spritsail barges were the dumb lighters - barges with no propulsive power. Many wooden lighters remained after the First World War but the number built of steel was increasing rapidly. It was these and the damage they might do coming alongside that the sailormen feared. Several spritsail barges kept a dog on board to discourage lightermen from lying too close.

But the lighters made a big contribution to the operation of the Port. Over seven thousand lighters were then in daily use and their total capacity was more than a million tons. Before the Second World War they distributed about ninety per cent of London's imports.

All Thames dumb barges are swim-headed - with bluff bows sloping inwards - and end in a budget, a sort of fixed rudder. The design is the result of centuries of evolution and is unlikely to be improved for tidal work. Apart from special craft, they range in capacity from 50 tons to (for grain) 750 tons. The most popular size in the 1920s loaded 160 tons.

Today practically all remaining Thames lighters are towed but, up to the outbreak of the Second World War, many of them - and especially those belonging to the smaller lighterage firms - were each 'driven' by a solitary lighterman. With the aid of his 'paddles' - oars more than twenty feet in length - he would row the laden craft upstream or down, using every eddy and set of tidal current to help progress.

These men were the cream of the tideway individualists. With an inherited background of river lore, for most of them came from generations of lightermen, they were prickly, proud and fiercely independent. Above all, their loyalty to their employers, even the more flinty-hearted ones, was proverbial. They all served an apprenticeship of from five to seven years and, after 'passing out', were licensed by their 400-year old guild, the Company of Watermen and Lightermen of the River Thames. In the 1920s, their jealously guarded status was usually signified, even when afloat, by the bowler hat. Certain unlicensed lightermen were permitted to move barges about in the enclosed docks and these were patronisingly referred to as 'barge pokers' by the licensed professionals.

Completely exposed to the weather, the latter would take their craft on a passage, say, from Rotherhithe to Tilbury, a distance of more than twenty miles, probably using the motive power of two successive ebb tides. Rain or blizzards would be ignored. Their greatest fear was of being marooned in fog, unable to get ashore. The only shelter was a cubby hole aft, equipped with a small stove but perhaps no fuel, and they usually had only sufficient rations for the passage. An extra day or two under these conditions meant real hardship.

A famous former lighterman whom I got to know well was the late Tom Scoulding. Elected Alderman and Mayor of West Ham Borough Council, appointed a Justice of the Peace and a representative of the Ministry of Transport* on the Board of the Port of London Authority, these and many other honours came to him in mellow old age.

When he was apprenticed lighterman in 1892, he worked a 72-hour week for 8s, and an all-night shift would bring him 1s 6d. When he was out of his time in 1898, he received the full wages of a lighterman; 6s for a 12-hour day spread over 14 hours of attendance, with another 4s for a 'short night' up to midnight, and another 2s if he worked a full night.

A passionate fighter for the then exploited port worker, Tom became a full-time trade union official in 1903. But before that his natural pugnacity had landed him in trouble with the police during a strike.

---

* In 1920 the Ministry of Transport succeeded to the powers and duties of the Board of Trade in relation to the PLA.

When he appeared in the dock at Grays Police Court the magistrate looked gloomily at him and then asked the Court to stand.

'I have,' the magistrate said, 'a most sad and solemn announcement to make.' (Tom was expecting a black cap to be produced.) 'I have received news,' he continued, 'of the death of our gracious Sovereign, Queen Victoria. Long live the King.'

Tom was acquitted.

I met other types of river workers during my early explorations, and many of them have been swept away by change. A river postman, the last of five generations of his family so employed, used to row daily from London Bridge to Limehouse and back again, carrying letters to ships in the Pool. Today their mail is either delivered to the wharf where they lie or, if they are moored in the stream, to their London agents.

Unemployed watermen would sometimes scull slowly along the tideline seeking 'bluey' - pieces of copper and brass which fetched good prices from scrap-metal dealers. The concrete barge hards at some of the wharves were kept clear by 'luters' who, protected by thigh boots, swept the accumulation of mud into the river at low water with enormous squeegees.

Another group of individualists were the PLA piermasters in charg.* of the floating piers (miniature landing stages for river craft) between Woolwich and Teddington.* Their duties were varied; at one moment they would be dealing with, say, fare-paying old ladies, and the next with tough tugmen or launch crews.

One rather abrasive character I recall used to boast of his 'tact' which, at close quarters, could be devastating. I was once official host to some important visitors touring the Port and I asked him to clear his pier as tactfully as possible of a party of noisy school children just landed with their teachers from a launch. I wanted to embark the guests in the PLA yacht with some decorum.

'Leave it to me, sir,' he replied. He took a deep breath and let fly: 'Now, you little bastards, if you aren't off this blee'n pier in one minute, God alone knows what I won't do to you.'

---

* Most former PLA river piers are now controlled by the GLC.

Within seconds he came back to me. 'They're gone,' he said with the satisfaction of a job well done.

Practically all the river men were more or less exposed to the weather; most of them by contemporary standards were poorly paid; few had security of employment. In spite of, or because of, these drawbacks, robust tideway humour was plentiful. In the days when I first knew them, only bucolic, slapstick fun brought out the deep belly laugh of these men. My friend, Charles Alexander, told me a story of his youth which aptly illustrates the sort of yarn then most relished.

He was serving as mate of a tug that was just stemming the falling tide and waiting for another few inches of head room before passing under Hammersmith Bridge. Gaping down on the tug from the bridge parapet was the usual London crowd, including a young policeman. The tug-master, a fiery little waterman with a red beard, let go a blast of his steam whistle almost in their faces. Back jumped the crowd and down to the tug's deck fell the constable's helmet.

He made the mistake of demanding the return of his helmet in a very peremptory way. After instructing the constable to address him as Captain, the skipper tried on the helmet, thereby snuffing himself to the roots of his beard. Then he gave the policeman a lecture on the iniquity of a mere landlubber poking his nose into the mystery of tidal affairs and concluded by tossing the helmet up to the bridge. His throw was nicely timed to coincide with the passing of an open-topped horse bus, and the constable was last seen sprinting after the vehicle.

But humour along the river - as elsewhere - has undergone a subtle change during the last few decades. There is a taste for the wisecrack, due no doubt to greater literacy, as well as to the influence of films, radio and, more recently, the all-pervasive television.

A colleague of mine was one day on Westminster Pier when a tug came upstream with a fault in its whistle which caused it to give out a continuous shriek. On being asked: 'What's he blowing for?', the Piermaster replied with a poker face: 'He's always in trouble. He hit London Bridge last month and Blackfriars Bridge last week. Now he's blowing for Westminster Bridge to get out of his way.'

This crack was a great success. It would hardly have raised a smile in the old days.

Even at fortune's lowest ebb, the people of the river maintained the blunt independence which has been and continues to be one of their chief characteristics. In those days, most dockers carried their lunch clapped between two plates tied in a bandanna. Once, I inadvertently sat on such a packet. At the sound of splintering, I sprang up. The elderly docker who owned it gathered together his broken fragments in the bandanna without a word. 'I'm fearfully sorry,' I ventured. He did not reply and I repeated my apology. Then he spoke in measured tones. 'If I,' he said, 'had a bloody canary as pretty as you and I wrung its blee'n neck, saying I was sorry wouldn't bring the poor little bugger back to life again.' He said no more, ignoring my offers of compensation.

chapter eight

Gravesend Reach

1

GRAVESEND REACH WAS the focal point of the Port's traffic. Here the launches of the Pilotage Service, the Port Health Doctors and HM Customs intercepted the deep-sea ships. At the tug buoys clustered my particular friends, Alexander's *Suns*, as well as craft of other towage companies including those of William Watkins, Ltd, the oldest towage company in the world. It was Watkins's famous paddler *Anglia* (known as 'Three Finger Jack' from the arrangement of her funnels) that towed Cleopatra's Needle from Ferrol in Spain to the Thames.

Along the Gravesend shore, opposite Tilbury, were the little bawleys - shrimpers which trawled in the outer estuary, most of them under sail, although even then some motors were being installed. Bawleys are said to have taken their name from the boilers on board in which the catch - the sweet brown Thames shrimp - was boiled on the homeward run. Still earning her living with this fleet was the *Ellen*, reputed to be some two hundred years old, but now vanished from the Thames.

2

On Gravesend's Royal Terrace Pier was the London headquarters of the Trinity House pilots under their picturesquely named Ruler of Pilots. These men, amongst the elite of the river hierarchy, were members of a wider semi-autonomous association but licensed by the Corporation of Trinity House, the Pilotage Authority for London as well as for a number of other British ports.*

Here were gathered in my day the Channel or sea pilots, (working from Gravesend outward both to the north and the south), compulsory river (Thames) and exempt river (Thames) pilots. The London District extended northward to Orfordness (the Sunk Light Vessel) and south to Selsey Bill (though pilotage was compulsory only as far as Dungeness).

For reasons of decentralisation and convenience, there were other smaller stations, in addition to Gravesend (claimed to be the largest and busiest in the world), further seaward for such as the North Channel pilots (from Sunk to Gravesend), etc.

A familiar sight at Gravesend was the little pilotage launches (not to be confused with the much larger cutters which cruised off the outer estuary) putting out from the pier, dealing out river or 'mud' pilots to inward-bound ships and taking off sea pilots; and vice versa for outward-bound vessels.

It was the River Thames - the 'mud' - pilots whom I came to know first. Not until much later, when I ventured into the outer estuary, did I meet some of the sea pilots operating downstream of Gravesend. In the growing economic depression of the 1930s, which was causing not a few master mariners to sign on before the mast, the comparatively high earnings of the pilots were a magnet to seafarers. But many of these pilots - most of the sea pilots, at any rate - had been lured from big ships by family ties, and family loyalties were often strained at each fresh contact with a fine vessel outward bound. It is a problem almost as old as seafaring. Moreover, some of the pilots spent many hours travelling and living out of suitcases, thus apparently getting the worst of both worlds.

In the early 1930s, ships' bridges were not festooned with the sort of electronic gadgetry that is routine today, so the skills of the pilots, although not lessened by these aids, were the more apparent. To the shape and bearings of each winding river reach, carried indelibly inside their heads, they brought local lore, especially valuable when visibility

---

* Besides being the Pilotage Authority for the London District, the Corporation of Trinity House continues to be responsible for navigational beacons, buoys, lights and seamarks in the Thames.

was just sufficient to allow vessels to be under way. The old river jibe about navigating by the bark of the piermaster's dog was perhaps far fetched, yet I know of more than one fog-bedevilled pilot to whom a sniff of the gas retorts at Beckton or the smell from the Silvertown Soap Works was as good as a fair-weather fix.

In the days before radar, when fog might hold up shipping for a tide or two, a sudden clearing would set every vessel under way in an effort to make good some of the lost time. Then the skills of the pilots were even more vital, with heavy inbound and outbound traffic mixed up on the same tide.

Tugmen, perhaps a little envious of the pilots' higher status, delighted in malicious tales about some of their foibles. According to them, the enormous appetite of one old mud pilot was so well known that as soon as he arrived on the ship's bridge a steward would be at his elbow with a luncheon or dinner menu. He would take the card and begin: 'Slow ahead both. I think I'll start with some *hors d'oeuvre*. Keep her on that buoy. Then some grilled plaice. Steady as she goes. After that I'd like the grill. Starboard a little . . .' and so on.

Another long and involved story was told of a dashing young pilot who was reputed to have landed one of his ships into the middle of a ploughed field. A Dutch skipper, terrified of his 'full ahead' technique, was alleged to have described him with mock enthusiasm as the greatest Englishman of all time. 'For,' he added, 'he has sunk more ships than your national hero, Nelson.'

3

One of my favourite Gravesend rendezvous was the Ruler of Pilot's office on Royal Terrace Pier. His windows commanded a view of the Port's most fascinating scenes: arrivals and departures to or from the landing stage at Tilbury across the river; the tidal panorama of international shipping, both inward and outward; the constant coming and going of official launches and watermen's skiffs; and an occasional spritsail barge. Back and forth through the melee like dignified dowagers moved the Tilbury-Gravesend ferry steamers.

This office was also a good starting point from which to learn something of the splendid drama and variety of ships' flags. New methods of communication will inevitably render them superfluous but when I first went exploring along the Thames they were the eloquent tongues of seamen speaking a language almost as old as sea carriage.

An ancient signal surviving from the past was the weft - a knotted ensign once used at sea as a distant signal (now abolished) in the International Code but today meaning in port that a customs officer is required on board. Vessels approaching Gravesend and needing to exchange a sea pilot for a river pilot or vice versa would blow one long and four short blasts (Morse code for TH, signifying Trinity House) and would indicate that a pilot was on board by flying flag H in the International Code. A plain red flag (B International) would warn that explosives were being handled. Plain yellow Q of the International Code (the Yellow Jack) told the Port Health Authority that the vessel was healthy and required free pratique, i.e., permission to make contact with the shore.

More heart-stirring to me was the Blue Peter, International Code P, flown at the foremast in port to signify that the ship's company should repair on board in readiness for sailing on the next tide.

The P & O always kept a medical officer on duty in one of its several ships in port together and that ship showed his presence by flying M International. And when the Orient Line vessels in Tilbury Docks were sealed and fumigated with deadly prussic acid gas at the end of each round voyage they always flew the Jolly Roger, its skull and crossbones serving as a warning to keep clear.

I got to know the house flags - those flags of identity usually worn at the mainmast on entering or leaving port. Many of them proved to be evocative of their companies' history.*

The P & O house flag, for instance, incorporated the blue and white of the Portuguese royal colours and the red and yellow of the Spanish ensign to commemorate the company's early trading connections. The Shaw Savill flag was once the national flag of New Zealand before those islands were ceded to Great Britain in 1840. The New Zealand Shipping

---

* A fine collection of house flags, presented by the late Daniel R. Bolt, an authority on the subject, to the Central Library of the London Borough of Tower Hamlets, is now on permanent loan to the National Maritime Museum, Greenwich.

Company's house flag included a separate pennant, surviving from the days when this line owned both sailing and steam ships. The pennant (the 'steam cornet') was added to distinguish steam from sail; not so easy when the steamer's funnel was obscured by clouds of canvas to which steam was then only a dubious auxiliary.

The Brocklebank flag was distinguished by being worn at the foremast instead of, as customary, at the mainmast. Various reasons for this deviation have been advanced; one of the most likely is that in the days of sail this company's two-masted brigs were sometimes under Letters of Marque and therefore flew a pennant at the mainmast.

But if there is much maritime history in the house flags of London's River, there are also many yarns about them which do not bear investigation; of disputed flags being nailed to the mast, and the rigging and stays being greased to prevent their being struck; of handkerchiefs being pinned over offending designs; of eccentric customs in old windship lines being immortalised in their house flags.

It was the late Ernest W. Green, a descendant of the original Green, partner of Wigram in founding their famous line of sailing ships, who demolished one popular myth. The story went that Wigram and Green's new house flag consisted of a white square with a red St George's Cross in the centre and that it was first worn by their new ship, *Sir Edward Paget* (1824). But when the vessel, outward bound, put in at Spithead, the Navy is alleged to have ordered the flag to be struck as it infringed the traditional Admiral's flag. Thereupon a midshipman in the *Sir Edward Paget* is said to have cut off the tails of his uniform coat and pinned a blue square over the centre of the cross.

Owing to a subsequent quarrel between Wigram and Green, after which a wall was built through the famous Blackwall Yard and priceless models were childishly sawn in half, the flag survived in two versions. In its original design it became the house flag of the Federal Line. The second version, with the cross superimposed on the square, appeared on the cap badges of George Green's School and on the caissons of the dry docks at Blackwall Yard.

Ernest Green would have none of the old story about the coat tails. He told me that while the origin of the flag was unknown, he was at least certain that Wigram and Green, both shipbuilders and experienced seamen steeped in Royal and Merchant Navy traditions, would never have made the elementary mistake of choosing an Admiral's flag for themselves.

4

Over the years, Gravesend itself became my principal hunting ground for Thamesiana. A suburb in the hinterland, it was all seaport along the river. One of the characters I came to know there was the late W. T. Starbuck. He had been apprenticed during the last century in his father's fleet of sailing colliers and fishing smacks. His family had handed down, from generation to generation, personal memories of the Press Gang stealing Gravesend fishermen, with the ruses and bribery used to escape their clutches; of record catches of fish; of complaints about God's bias in the matter of fair winds for rival French smacks while the Gravesend men - 'His own people - were becalmed inshore. All that remained of those simple clannish Gravesend days was the sight of nets triced up to dry at the mastheads of the bawley boats lying on the foreshore mud.

Then there was old William Warner, last of the Gravesend skiff builders. His shop, all of timber, was over four hundred years old. Most of his craft were built of oak, copper fastened throughout; the timbers and knees were cut from crooks grown to the required shape and fastened to the planking by a process known as 'joggling'. If you wanted a cheaper boat of steamed timbers, he would build it for you, but with undisguised disapproval.

The Gravesend skiffs were usually manned by two watermen, and William Warner recalled that in the great days of sail the first crew to get their skiff alongside a homeward bounder won the contract for docking her when she arrived in the Port. The watermen would beg a tow for their skiff from an outward-bound vessel, sometimes going as far afield as the Royal Sovereign Lightship,* and making the return passage with their prize. For their docking services, the two men would share a fee of not more than thirty shillings, although there was usually a bonus of saleable old rope and fat from the galley.

Then there was the marine collection of that colourful Gravesend character, the late Captain Long John Silver. A man of lively spirit, he could be said to have inhabited two worlds. One, spanning Monday to

---

* Now replaced by an automatic light.

Friday, was set in London where he was the apparently conventional business man. The other, in which he was able to indulge his unorthodox personal tastes, was his week-end home on the Gravesend waterfront.

He had named this The Look-Out and fitted it up inside as a ship. It had a gangway, a bridge ladder, a bridge complete with engine-room telegraph, wheel, binnacle, port and starboard lights and a voice tube to the steward who served seamanlike drinks to the guests peering at the river scene through the bridge screen. John was nothing if not hospitable. As he aged, the part of Long John Silver grew on him, until it became difficult to separate the man from the image. But behind his play-acting was a shrewd brain, and his private nautical museum at The Look-Out was an expression of his love for ships and the sea.

The collection, set out in glass cases, included hundreds of models, marine curiosities, relics, etc., flanked by more than eighty merchant ship figureheads. Some of the heads were famous (one reputed to have been carved by Grinling Gibbons); others had come from undistinguished ships of which little more than the name had survived; in some cases the origin was unknown.

When the lease of The Look-Out expired in 1952, the Silver Collection was presented to the National Maritime Museum. The figureheads are now displayed in the hold of the *Cutty Sark* at Greenwich. John Silver did not long survive the dispersal of his collection. He died, a worthy, if eccentric, Thamesman, in 1959.

In the 1930s, there were still many survivors of the long line of taverns which once monopolised practically the whole of the Gravesend shore-line. The Three Daws, which began life as five cottages built at the end of the fifteenth century, has held a continuous licence since 1565. It is such a labyrinth of rooms, passages and stairways that, during the Napoleonic Wars, the Admiralty issued a special order whereby the Press Gang was to be reinforced when raiding the tavern as so many men escaped through its warren.

The Royal Clarendon Hotel is too grand to be described as a tavern. Its air of comfortable superiority is inherited from the seventeenth century when it was built as headquarters for the Lord High Admiral, James, Duke of York (later James II), whose first wife was the daughter of Edward Hyde, Earl of Clarendon. The portico needs only a fall of snow and a waiting coach and four to recreate the traditional Christmas card setting. But the atmosphere inside is modern and belongs to any prosperous provincial town in England. Somehow, it is not of the tideway.

Next to William Warner's old boatyard was the derelict building (demolished in 1954), formerly the Amsterdam tavern, whose history went back to the sixteenth century. It was claimed as the original of the capstan chanty, *A-roving*, which begins: 'In Amsterdam there dwelt a maid'. But research seems to prove that the Amsterdam in the song was, in fact, the Dutch seaport.

A look at the ramshackle building which had once been the Spread Eagle led me to the story of the Waterside Mission Church. It began in the latter half of the last century when the Vicar of Holy Trinity, Gravesend, started to visit the large waterborne community living in hulks and, later, the emigrants on board ships in the river waiting to sail. From this small beginning based on the old Spread Eagle building came St Andrew's Mission Church, consecrated on St Andrew's Day 1871.

The Church became even more closely associated with the emigrant ships and it was the cutom to ring the bells whenever such a vessel sailed from Gravesend Reach. Thus many settlers left the country with a last memory of English church bells echoing in their ears. For years the church tower leaned out of the true and as such was a well-known landmark for shipping. Its tilt worsened until it became unsafe for bell ringing. But it was rebuilt in 1949 and its bells could be rung again. Now the church itself has been closed.

Off the Denton district of Gravesend before the Second World War lay the training vessel, HMS *Cornwall*, a 74-gun teak ship built at Bombay in the early nineteenth century and claimed to be one of the oldest ships afloat. She came to an untimely end by air attack and all I saw of her after the war were her bones on the river foreshore. Like many ships built of teak, her ribs and beams had been earmarked for furniture.

chapter nine

To the Estuary

1

BELOW GRAVESEND, BETWEEN the world wars, cranes and cargoes, landings and warehousing, thinned out until there were only a few specialised trades such as crude oil, cement and explosives. For me, this was where the Estuary virtually began, and the river suddenly became a highway to the world's far places. Here I started an affair of the heart - especially with the outer estuary - which has never faded.

The fairway downstream of Gravesend (leading into Lower Hope Reach) is lit by navigational buoys, flashing or occulting in their ordered rhythm. During the thirties there was precious little else to see along the lonely Essex and Kent saltings until one came to the ruins of Cliffe Fort (on the Kent side) built by Henry VIII and rebuilt by Gordon of Khartoum. During the Second World War, I sometimes visited the fort, by then a comfortable Naval outpost with a high standard of hospitality.

Downstream are Lower Hope Point and Sea Reach where the river again widens, ceases its meanderings and turns uncompromisingly eastward for a straight run to the outer estuary. In severe easterly gales, the Point provides the last substantial shelter outward bound, and, before the Second World War when spritsails were a feature of the river, a sure sign of unpleasant conditions 'outside' would be a fleet of sailing barges anchored in its lee.

On the Essex shore of Sea Reach is London's oil port, with the shining tanks of a huge storage depot and the writhing pipes and towering structures of two refineries.* A seamark of this area is Holehaven Creek, the western boundary of Canvey Island. Here, in contrast, is the Lobster

---

* Further developments in this area, with its essential deep-water frontage, are planned.

Smack tavern, a survivor from the days when Norwegian lobstermen moored in the creek until prices were high enough to be worth taking the catch up to Billingsgate Market.

The Old Lady of Canvey (more properly, Scar Beacon) and the Chapman Light were important leading marks during the last century when vessels coming in from sea would keep the two in line for a safe approach into the narrowing river. The Old Lady, which disappeared in the early 1960s, was a land-based beacon, so called because of its distinctive shape. The Chapman Light off Canvey Island marked the Chapman Sand and, until it was removed in 1956, was the Port's only manned lighthouse. It has been replaced by a buoy.

2

Downstream of Canvey Island is old Leigh-on-Sea which has been continuously linked with the fishing industry since the thirteenth century. It was for long the home of a guild of Thames pilots who, it is said, united with the pilots of Deptford to become the Corporation of Trinity House in the first years of the sixteenth century. The Pilgrim Fathers' *Mayflower*, and a particularly active press gang as well as the catching and preparation of cockles are also associated with Leigh.

Eastward is Southend-on-Sea, the playground of East London. Cars and coaches bring crowds of visitors down every weekend – on fine summer evenings as well – and the town has adjusted its image accordingly. The thousands of London commuters who live there can, during the summer, approach the sea only with determination.

Nevertheless, I have known moments of enchantment there when the effects of light and shade on the shoreline and on passing vessels transcended the unlovely bustle of tripperism. For the inhabitants it is a matter of pride that the twice-daily ebb of the estuary waters lays bare so wide an expanse that Southend Pier* is claimed to be the longest in the world (over a mile).

---

* Badly damaged by fire in 1976. Official policy is that the damage should be repaired, but owing to financial stringency, no date can be given. The pier railway has been closed for repair and its future is in doubt. However, the pier walkway is open and will no doubt continue to attract Londoners with a taste for sea breezes.

Plate 9:

Spritsail barges awaiting work tied up at Woolwich Buoys. The crews, existing on 'subs' from the owners, referred to the moorings as 'Starvation Buoys'. *Photo: Stanley.*

Plate 10

The Central Electricity Generating Board 'flatiron' collier, *Fulham X*, in the Upper Pool. These vessels, with cargoes of coal for up-river gas works and generating stations, are built with a very low superstructure to enable them to pass under Thames bridges. The depth of water under the bridge arch and the clearance above are sometimes little more than a matter of inches. *Photo: Port of London Authority.*

Beyond Southend are Shoeburyness and Foulness Island. Offshore is the desolation of the shoaling Maplin Sands that at low tide are revealed as a great sandbank some twenty miles long.

This Essex coast, with its network of creeks, its history of recurring floods, its lonely farmsteads and sparsely scattered inhabitants, has its own triste charm. But Foulness and the Maplins have been the subject of furious debate as a possible site for Britain's third international airport. More pertinent to my story are proposals that part of the area should be developed as a large new London outport comparable to the giant complexes at Rotterdam, Antwerp and other Continental competitors of the Port of London.

The scheme would enlarge the responsibilities of the PLA because, in 1964, the Port's seaward limit was extended by twenty-two miles, taking in the Essex coast-line as far north as Foulness Point and then moving northeast to include most of the outer estuary.

The extension has already enabled the Authority to look at the future and create another fairway from the sea, intended mainly for deeper-draughted vessels. The new route bears the euphonious name of Knock John Channel.

3

On the south side of the river, opposite Southend, is BP's Kent Refinery on the Isle of Grain. Although its jetties and installations front the Medway, its gleaming towers, sometimes given an unreal mirage-like quality by low-lying mist, have become a noted Thames landmark since the Refinery was opened in 1953.

Grain is no longer an island. At one time it was separated from the mainland by Yantlet Creek which has since been blocked by a causeway. As the downstream limit of those earlier conservators of the Thames - the City of London Corporation and its successor, the Thames Conservancy - the Creek was a signpost to the Port. This link is commemorated by the use of Yantlet's name for a vital part of the dredged ship channel, extending from Sea Reach Number One Buoy (near which the *Nore* Light Vessel formerly lay) up to Lower Hope Point; a stretch on which dredging had been started by the Thames Conservancy, then extended and maintained by the PLA, and recently dredged to still greater depths to suit the bigger crude oil tankers that discharge at Thameshaven and Coryton.

The Isle of Sheppey, on its heights above Sheerness, has the Abbey Church of Minster, a seamen's landmark for hundreds of years. The vane of the church has a horse's head instead of the conventional arrow and is surmounted by the figure of a horse. This commemorates the story of Sir Robert de Shurland, a twelfth-century Lord of Sheppey, who, according to the *Ingoldsby Legends*, died as the result of a contemptuous kick at the skull of his faithful horse, Grey Dolphin, which he had callously killed some three years earlier. His boots were in poor shape and one of the skull's teeth pierced his big toe and, although it was amputated, he soon died.

My copy of the *Legends* has a footnote stating that when the tomb was opened during repairs to the Church, Shurland's skeleton was found to lack one of the big toes.

4

It was the *Ingoldsby Legends* – 'Well, there the twins stand on the edge of the land to warn mariners off from the Columbine Sand' – that encouraged my sortie to the Reculvers, then outside the Port limits but very much a part of the Thames story. At one time, seamen used to show their reverence for the twins by lowering topsails as they passed.

Early in the nineteenth century, Trinity House had been just in time to save the towers (all that remain of the ancient Saxon Reculver Church near Herne Bay) from further erosion by the sea, as well as from local vandals. Such was the importance given to this navigational mark.

Also preserved is a substantial part of the Roman fort there, one of those maintained by the Count of the Saxon Shore, a system of coastal defences stretching from Norfolk down to the Isle of Wight. This fort was intended to safeguard the now dried-up valley of the Wantsum which was once navigable between the Thames and the English Channel near Sandwich when Thanet was an island in fact as well as in name.

5

As for the estuary itself, I came to know its many moods, taking every opportunity to savour its illimitable skies, misty horizons, ships and men. There were boisterous, cloud-strewn days with the wind whipping the water to turbulence, and other days when the grip of winter merged sea and sky into a cold grey wilderness; moonlit nights when evanescent shadows suggested new and eerie dimensions; or nights so black that, for me, the shifting lights on land and sea created only patterns of confusion; and summer dawns when the estuary and its ships took on fresh life with the sunrise.

The tidal traffic which, in the upper reaches, was so concentrated that ships seemed almost to jostle each other, out here was dispersed into distant pillars of smoke; or, nearer at hand, so dwarfed that Cunarders could look like coasters. Most of the big ships outward bound - P & O, Orient, Cunard, Ellerman's, Union-Castle and others - were heading down Channel for the open sea. Smaller vessels - north-east coast colliers, ships making for Denmark, Norway and the Baltic - were set on a north-easterly course. Or I might see local work craft - bawleys trawling for shrimps; pleasure seekers - summer yachts out of Westcliff and Southend, or 'Butterfly' steamers on day trips. In contrast, there was sometimes a lone warship making for the Medway and Chatham Dockyard.

Geologically speaking, it is not long since England was joined to Europe, not only to part of what is now France but also, by a marshy plain, to the Netherlands and Germany. Then the Thames was probably a tributary of a Continental river system which flowed into a northern sea somewhere between - or beyond - the Wash in Norfolk and Texel Island in the Friesian group. About eight thousand years ago, a great subsidence is thought to have occurred, when the tide broke through to merge and enlarge what are now known as the North Sea and the English Channel, and so made Britain an island. The shallow sea thus created has since been a testing ground for our seamen; such men as Drake and Cook learnt their skills here.

The Thames Estuary is a playground for the tides, some 250 square miles in extent and projecting more than thirty miles seaward of Southend. It is continually moulded and remoulded by the swirling currents. Its

tortuous and constantly shifting channels have been a significant factor
in London's long immunity from seaborne invasion. The North Sea tides
sweeping into and out of the estuary have formed channels radiating out
from the Nore Sand, somewhat like the fingers of a hand but more or
less parallel in a north-easterly direction. These channels are separated
from each other by shallower ground; banks of sand or mud, some of
which dry out at low water. The three main approach channels are the
West Swin, Barrow Deep and Black Deep, and through them pulse the
principal tides affecting the Thames.

Much smaller in volume are the tides which enter the estuary from
the Strait of Dover. These create further complications by cutting
subsidiary channels across the estuary sands and excavating what are
known as swatchways, some of varying navigational value since they
have tended either to deepen or to silt up.

Generally speaking, deep-sea ships kept to the main channels. For
those bound through the Dover Strait, the main route to or from the
Thames Estuary was the well-surveyed and buoyed Edinburgh Channel
(named after that earlier Duke of Edinburgh who was Queen Victoria's
son and Master of Trinity House). It has had a long history of fluctuating
widths and depths but it has served the Port of London well; only in the
past few years had it become necessary to supplement it by the new
Knock John Channel.

The smaller swatchways and numerous gutters and tidal creeks were
formerly almost the exclusive haunt of spritsail barges because of their
remarkably shallow draught and the skippers' unrivalled knowledge of
these waters.

The names of the channels and swatchways are part of the poetry of the
Thames - the Warp, the Swin, Oaze Deep, Black Deep, the Wallet and
others. Just as evocative are the names of sands and shoals - Gunfleet,
Shingles, Shivering Sand, Maplin, etc.

Some of the names betray their Old English or Scandinavian origins.
'Gore', as in Havengore, is from an Old English word meaning a
triangular spit of land. Girdler is a corruption of griddle which the sand
so named resembles in shape. Shivering Sand relates to a splinter, as
when a glass is shivered. Maplin once had an old causeway across its
sands from Wakering Stairs to Foulness Island and this was marked with
upended besoms that gave it the name of Broomway. But brooms made

from twigs were sometimes called mapples, which suggests a likely enough connection.

Then there were the lights; buoys kicking up their heels in the estuary popple, some of them whistling or ringing a warning fog bell; and the more sedate swinging and dipping lightships. The rhythm of their flashing or occulting was all recorded on the charts so that even an occasional passenger in these waters could learn to recognise some of them. Their names were yet another litany of seafaring - *Mouse, Tongue, Edinburgh, Girdler, Mid Barrow, Nore* ...

6

On a two-day voyage in the Trinity House steamer *Alert* (lost, alas, at the Arromanches landings) I visited some of the estuary lights.

Our first call was the *Nore* lightship, built in 1839, and the oldest such vessel then in service. Constructed of wood, she had an oil lantern which revolved by means of a weight and pulley. Her foghorn and most of her equipment were operated manually. Low deck beams were festooned with nets of apples, hocks of bacon and other provender, for the crew brought their own food for each tour of duty; a month in the case of the Master and two months for each of the five hands. Hammocks instead of bunks added to the impression of stepping back into history.

Ships straying towards danger were warned by an old-fashioned cannon fired by gunpowder. Sponge and rammer for this weapon were in contrast to the crew's only concession to the twentieth century - a radio receiving set. Judged by modern standards she was out-of-date, but very much in keeping with the Nore Sand's tradition of 'wooden walls', old-style naval pageantry, muzzle-to-muzzle sea fights and ghosts of the Nore Mutineers.*

For the rest of the day we recharged and refurbished some of the estuary light buoys and spent the night anchored in a swatchway between two

---

The *Nore* lightship has now joined the collection of historic vessels in St Katharine Yacht Haven.

sheltering banks. Next morning we visited the *Tongue* lightship. By comparison with the old *Nore*, the *Tongue* was several generations ahead. She was a steel ship equipped with a battery of semi-diesels which produced power to operate pumps, windlasses, fog diaphones and dioptric light plant. Radio telephone transmitter and receiver, well-planned cabins and modern storerooms added to the safety and comfort of the crew and pointed the road from the early days of the *Nore*.

7

Another inter-port ship in which I sailed the estuary before the last war was the *G. W. Humphreys*,* one of the then LCC fleet of sludge vessels† which, on every tide, took the solid matter from the sewage treatment centres at Beckton and Crossness out to the dumping ground in the Black Deep. Not for nothing were these ships known to rivermen as the 'LCC yachts'. Immaculate and well found in every way, with excellent crew accommodation, they were run as smartly as P & O liners and, in fact, some of their masters had been officers of that line.

As tide time was difficult, I slept on board the night before my trip, and I was impressed next morning when the steward knocked at my cabin door and grandly announced: 'Your bath's ready, sir'.

Only during the brief loading period, when about 1,500 tons were pumped into the ship's four tanks, was there any odour. Continuous samples for the LCC chemist were taken during the operation. The working fleet of three vessels dumped some 47,000 tons per week and covered on each round trip more than a hundred miles.

Near the dumping ground, we made our number and saw the answering pennant hoisted by the *Edinburgh* lightship; a record would be taken to show that *G. W. Humphreys* was dumping in the approved area.

---

* The *G. W. Humphreys* was lost by enemy action in October, 1940.
† Now operated by the Thames Water Authority.

The boatswain and his hands appeared on deck, ready to manipulate the eight valves which would allow the buoyancy of the ship to force out the cargo. Ahead of us a vast cloud of sea birds swooped and wheeled in anticipation of the feast to come; like the swans up river, their intelligence service was faultless. East of **D2** Buoy, the bridge telegraph rang for half speed and we turned sharply to port. Our creamy wake abruptly changed to black with the discharged sludge, the *G. W. Humphreys* rose higher in the water – and the birds dropped down.

8

I sometimes went out with **PLA** Hydrographic Surveyors, ceaselessly charting the shifting shoals and deeps of the inner estuary (before the seaward limit was extended). With echo sounder, sextant, a profound knowledge of the tides and, later, electronic aids developed during and after the war, they were responsible for the thousands of tiny figures which recorded low-water depths on Admiralty and **PLA** charts of the river.

These surveyors also plotted the exact positions at which dredgers were to be stationed and mooring buoys placed; co-operated with Trinity House in checking the position of navigation buoys; and provided information for the raising of wrecks, etc.

To get a line of soundings, the surveyors aboard the hydrographic launch made careful preparation. Sextants were used to plot angles from the starting point of the run to recognised landmarks. Tidal readings were timed with a synchronised watch. At the landward end of the run, their dinghy was anchored and its position, too, checked by sextant.

Then with the signal hoist 'O N E' (engaged on surveying operations) at the halliard peak, the launch would begin her run. A surveying officer on the foredeck would make continuous checks on the landmarks, while another officer in the anchored boat would check the course of the run. The master of the launch would himself take the wheel and his steering would be to a hairsbreadth accuracy. And during the run, the echo sounder would record the contours of the river bed. Then would follow hours of plotting to add a few tiny figures to the thousands of soundings shown on estuary charts.

9

The people of the river were among the toughest and most enduring of London's citizens, but perhaps those whose duties took them out to sea, still within the estuary but beyond what was then the official port limit, were just that much tougher.

One, in particular, remains clear in my memory; Captain Jewiss, master of the Tilbury Contracting and Dredging Company's tug, *Danube III* - perhaps because I, too, was to have later wartime adventures in that same craft.

The *Danubes*, stout ocean-going tugs and then the largest in the Port of London, made regular runs out to the Black Deep towing self-emptying hoppers laden with some 2,000 tons of 'spoil' from PLA dredgers. Like the sludge vessels, they were under way on practically every tide and they, too, had to make their number to the *Edinburgh* light vessel.

I joined Jewiss's craft at Gravesend. He settled me in a corner of the wheelhouse and began to tell me about his adventures. Below the Lower Hope, a north-east gale was raising something more than a popple in Sea Reach and we prepared for trouble by letting out more towing wire. As soon as this was done, the hopper's crew disappeared below - they knew what was coming.

Plugging seaward again, the old skipper continued his long and rambling tale of how he had been given a temporary RNR commission and had taken a *Danube* out to the Dardanelles during the First World War. By the time he was well into his yarn, we were climbing slowly up huge oncoming waves, tobogganing down the reverse slope, scooping up and flinging mast high a lot of the estuary, and emerging with a running nose to repeat the performance. Spray flicked at the bridge screen like dried peas, and vicious squalls shook the wheelhouse doors. But the skipper's flow of words kept time with the reassuring beat of the engines.

When the dumping ground at D2 Buoy came in sight, he brought his well-embroidered yarn to its climax, punctuating his denouement - the sinking of an enemy submarine - with a long and a short blast on the steam whistle to summon the hopper's crew to manipulate the 'doors' of their craft. They wasted no time on deck, for, washed by every wave, it was a very dangerous place.

With the now unladen hopper high in the water and skittering about like a balloon, we turned back and, the wind now astern, found life much more comfortable. As the lights started to spring up around us in the gathering dusk, the old man eased off the wheel and chuckled. 'Lemme tell you about the time . . .' he began. I have forgotten his next yarn, but I do remember that he obviously enjoyed that trip almost as much as I did.

SECTION TWO

THE EMBATTLED PORT

Title illustration:

A German Heinkel 111 over the Port during a daylight raid. The Isle of Dogs, then the heart of London's dockland, is clearly marked by the river's great loop through Limehouse, Greenwich and Blackwall Reaches. The West India and Millwall Docks can be seen centre right. *Photo: Imperial War Museum.*

chapter ten

The Loom of War

1

BY THE MIDDLE 1930s, the crooked cross of the Nazis was becoming a familiar sight in the Port; the custom was for German ships to fly a black swastika in a white circle forward and their national colours aft. 'Strength Through Joy' liners brought their arrogant young passengers who sometimes used them as floating hotels while sightseeing. These liners generally moored in the river, but German cargo ships berthed in the docks or at riverside wharves. More than once independent-minded London dockers had set-to's with the German crews when they flaunted their adulation of the Fuehrer.

Looking back to those days when we seemed to be lulled by wishful thinking, it appeared that a clash, if it did come, would surely be between Nazi Germany and Soviet Russia. So we Thamesmen continued to take pride in our non-partisan port as the swastika passed the hammer and sickle on the tide. And German visitors were still welcomed to the Port of London and shown over vital dock installations.

After Munich some of us learnt of the plans prepared for a local Port Emergency Committee, representing all its varied interests (virtually the PLA but enlarged and with much wider powers), to take over if war came and ensure the continuity of the Port's essential operations. (In the event, the Port Authority, under the Chairmanship of Lord Ritchie, and later the Rt Hon. Thomas Wiles, PC, continued to function in a muted role as well as being part of this Port Emergency Committee whose chief Executive was Mr J. D. [later Sir Douglas] Ritchie, MC.) Another post-Munich measure was the organisation and training of a River Emergency Service of volunteer yachtsmen in readiness for wartime mobilisation.

There were plenty of apprehensive faces about, but the dockers never

lost their Cockney sense of humour. One of them, shovelling a load of
sand into a quayside ARP bin, had his leg pulled by his mates. Straight-
faced, he replied: 'This is work of national importance. The Camel
Corps is going to embark here, and a bit of desert will make 'em feel at
home.'

2

Most of my PLA contemporaries with a taste for adventure
enrolled in the Dock Operating Companies of the Royal Engineers. I
sent my name in for the newly formed Royal Naval Volunteer Supple-
mentary Reserve. My qualifications for the Navy were meagre - some
crewing experience aboard inshore yachts and sailing dinghies, and many
hours spent in ships and craft of the tideway. I was also taking courses
in navigation and signalling. But I had strong doubts that their Lordships
at the Admiralty would consider me a worthwhile candidate.

Then the PLA River Superintendent sent for me. An expert on tidal
flow and techniques of river control, he was profoundly bored by office
routine. Sometimes he would take evasive action by thrusting a file into
the hands of his assistant and say grandly: 'Let this matter take its normal
course whatever that course may be.' When I reported to him he was
at his most impressive. 'I have been ordered,' he said, 'to organise the
Thames and Medway Examination Service for the Admiralty and I am
compiling a list of suitable people with local knowledge who would
receive commissions in the event of war. I think you could be very useful
in boarding and checking neutral ships. Would you like to come in?'

The Thames and Medway! And I had been thinking of - perhaps -
an armed merchant cruiser in distant waters or at least a gunboat on
some foreign station. So I told him of my application to join the RNVSR.

'If war comes,' he said pontifically, 'it will probably be over before
the RNVSR is called up for training. I strongly advise you to take a bird
in the hand.' He was so convincing that I gladly agreed to join his
organisation.

Wearing his other hat as a commander (later captain) RNR, the River
Superintendent called two or three meetings of his future officers so that
we could get to know one another and learn something about our duties
in case of war.

The senior officer afloat was to be Commander J. R. Stenhouse, DSO,

OBE, DSC, *Croix de Guerre*, RD, RNR. Few knew that the white ribbon among his decorations represented the comparatively rare Polar Medal. He was an old Cape Horner and had served his apprenticeship to the sea in square-rig; he had been with Shackleton's Antarctic Expedition in 1914; had commanded a 'Q' ship under sail during the First World War; and was captain of the Royal Research Ship *Discovery* for five years, two of which had been spent on sub-Antarctic expeditions.

During one of our informal talks, Stenhouse pinned me down regarding my qualifications for a wartime job at sea. As I brought out my scanty ragbag of nautical experience, he looked more and more depressed. In desperation, I mentioned my interest in the old sailing ships and my theoretical knowledge of square-rig manoeuvres. He fairly beamed at me and clapped me on the shoulder, saying: 'Good! I'll make a seaman of you yet.'

Our vessels would consist of three large ocean-going tugs and two smaller craft tugs, all based on Sheerness where the Medway joins the Thames. The River Superintendent explained to us that the Examination Service is the first defensive measure put into operation by the Royal Navy at the approach of war. Our duties as Examination Officers would be complicated and there would be co-operation with Naval Intelligence regarding contraband cargo and enemy agents. Our ships would be armed and would mainly act as the Port's guard ships, halting and investigating all suspicious vessels entering the estuary, with particular attention to the possibilities of concealed torpedo tubes and mines. Only when we were satisfied as to the bona fides of a ship inward bound would we give her a coded flag signal to carry her past the guns of the forts guarding the Port entrance.

3

The last notable peacetime event in the Port was the berthing of the new Cunarder, *Mauretania*, in the Royal Docks on 6 August 1939. With a gross registered tonnage of 35,674, she was then the largest vessel to berth in a London dock; a vindication of early PLA planning. Yet, even while the ship was sliding into the King George V Lock, with so little room to spare that she might have been made to measure, the sky seemed full of menace. Could she ever become a regular visitor as intended? In fact, it proved to be her sole appearance in the Royal Docks.

4

By the end of August, it had become clear that war was inevitable. The normal work of the Port paused briefly and then abruptly changed course. Ships earmarked for Government service hurriedly shed cargo and peacetime fittings, ready to depart for secret destinations. The new Port Emergency Committee put passive defence plans into effect; fire, gas-cleansing and first-aid stations were manned; blackout and stringent security measures were enforced. An admiral and his staff moved into the PLA building ready to assume naval command of the Port. On both banks of the lower river guns were installed.

Port workers of all types who were reservists left for the Forces. The streets of dockland suddenly quietened as schoolchildren were evacuated, many of them by river in the 'Butterfly' steamers. From Teddington to the sea, the Port held its breath ready for the worst.

By chance, I was able to observe most of the Port's activities during the long struggle, but I began the war in the area I loved best - the outer estuary. On 31 August I was commissioned Sub-Lieutenant, RNVR, and ordered to report to the Extended Defences Officer at Sheerness. During a nightmarish journey (railway routines were completely disorganised by troop trains) I had time to reflect that this was the end of another epoch. Perhaps, I speculated sombrely, it was also the end of the Port of London as I had known it.

chapter eleven

The Estuary on Guard

1

NOT FOR NOTHING was wartime Sheerness known to the Navy as Sheernasty. Flanked by a dreary marsh, it had a street of shops, a railway station, a cinema, two or three hotels, several pubs and a colourless slum flanking the semi-derelict dockyard which had been hurriedly brought back to active life. Past the docks flowed the Medway to join the Thames off Garrison Point, a tall signal tower which was to become very much a part of my new life.

On 1 September, I reported to a tired, grey-faced Commander, RN, on the staff of the Extended Defences Officer. He told me that our five tugs were being fitted out - three big sea-going *Danubes* (numbers III, V and VI) and two small lighterage tugs. He hoped to have the latter ready that night and ordered me to report back late in the afternoon.

In due course, I stood on Cornwallis Pier and looked at my ship, HMS *Kite*. She was a pathetic little coal-fired steam tug, and her still-tacky coat of 'crabfat' grey did not hide the scars and wrinkles of a long life nearing its end. The six-pounder gun on the after deck looked as if it would be wagged when she was pleased, and a minute searchlight on the foredeck was related to a Victorian policeman's bull's-eye lantern. A very large white ensign, too big for her ensign staff, draped her loins like a horse blanket.

About a score of Fleet Reserve seamen fairly swarmed over her, trying to fit themselves into accommodation designed for her civilian crew of no more than four or five. She was commanded by an elderly, completely unruffled lieutenant, RN.

Our orders were in keeping with the ship. We were to patrol between the *Nore* lightship and the South Shoebury buoy, put our searchlight on an occasional inward-bound ship and look vaguely belligerent. But no action was to be taken until we had received further orders. It was quite dark by the time we let go; a few minutes later I discovered that there were no charts, no ammunition, no drinking water and no food on board.

As we steamed away from the blacked-out dockyard, our consort, HMS *Keverne*, a slightly larger version of our craft, followed us. Navigation lights were either extinguished or dimmed, and the estuary which I had known as a sort of marine Piccadilly Circus was now a black nothingness. We found the *Nore* by making rather unseamanlike casts in the dark. Having found it, and with no certain ideas about the site of the South Shoebury buoy, we were reluctant to part with the lightship and steamed round it all night. We saw no other ships.

At dawn, I discovered that my new uniform bore generous patches of grey paint. And we decided that we were too thirsty to be hungry.

Then the leading stoker who was chief engineer came on the bridge and told the captain that the tug was making water fast and that we should have to return to harbour. To give weight to this verdict, he had donned a huge, cork life-saving jacket. A signal to Garrison Point eventually brought us the order to return forthwith. In the meantime, we had sighted HMS *Keverne*, broken down and at anchor in the Medway Channel. When I landed I was told to report to the Britannia Hotel. We never saw HMS *Kite* again.

No more ignominious introduction to the Navy could have been devised, but, luckily for morale, Stenhouse and the rest of our party had arrived at the hotel and they greeted the story of my adventures with hoots of laughter.

In the evening (2 September), now bathed, watered and fed, and with my uniform smelling of turpentine, I went afloat again. This time I moved from the ridiculous to the sublime, for I now boarded HMS *Salamander*, a fleet minesweeper filling the gap until our own ships were ready.

Vessels of this class, known colloquially as 'smoky Joes', were slightly smaller than the average destroyer of those days, but after my sufferings in *Kite* this temporary examination ship seemed to fall little short of a luxury liner. Clean, efficient and quietly carrying on the traditional peacetime ceremony of the Royal Navy, with boatswain's pipes and crew manning stations for leaving harbour, I gave her my impressionable heart.

All night we plodded up and down the estuary; as yet we had no real duties and were still only showing the flag. At dawn *Salamander* was ordered to return to Sheerness and, somewhat reluctantly, I landed and reported to the Extended Defences Officer. At 11.15 a.m. the Prime Minister announced that we were at war with Germany and, as he finished, the air raid sirens began to wail. As official histories have recorded, it was a false alarm.

In the afternoon I was taken by picket boat to BV10, an ungainly-looking telephone ship at permanent moorings in the river which was to be one of our links with Sheerness. There I was picked up by *Danube III* in which I joined Commander Stenhouse as junior Examination Officer. There were two other officers. Lieutenant Gilroy, RNR, was a fine seaman who taught me much, not least what was meant by being a good shipmate. Lieutenant Sullivan, RN, had been commissioned from the lower deck and, between the wars, had been third officer in the training ship *Worcester*. His Irish 'bulls' were a constant joy. For instance, he vaguely remembered that there had been WAACs in the First World War, and these he called Wags. In the Second World War, he differentiated between the Women's services by calling them Sailor Wags, Soldier Wags and Airman Wags.

2

      Danube III was the only coal burner in the flotilla and consequently difficult to keep as clean as her oil-fired sisters. After *Kite*, however, I was not disposed to look down my nose. The *Danubes* were each about 120 feet long and drew some sixteen feet. Over the engine-room skylight a platform had been welded as mounting for a twelve-pounder gun. Between the funnel and the bridge another structure had been added for a searchlight. On the roof of the wheelhouse were twin Lewis guns. She was now very different from the ship in which I had listened to Captain Jewiss's yarns.

    I was much impressed by our twelve-pounder until Sullivan told me it had been used in the Boer War and again in the First World War.

    The accommodation in *Danube III* (like her sisters) had been ample for the civilian crew of about twelve, but she was flooded with Fleet Reserve seamen; our total complement, including four officers, was twenty-two. The officers and four petty officers lived forward; naval custom decreed that the large saloon on the main deck should be halved to provide a wardroom for the former and quarters for the latter. To three sleeping cabins on the deck below we added a makeshift bunk in the bathroom. The eight of us were uncomfortably cramped.

But much worse was the seamen's mess deck, a gloomy cavern aft where they lived, sleeping packed tight in bunks and hammocks. One of our two Royal Marine signalmen had found room for his hammock only by slinging it above the mess deck stove whence he emerged for his watches rosy faced and half grilled.

3

The estuary was already vastly changed from the waters in which I had voyaged before the war. Most of the great liners had disappeared, all the pleasure craft had gone; even in sunshine, the estuary seemed a bleak and somewhat hostile place. Apart from an occasional spritsail barge, single ships soon became rare visitors as the convoy system got under way. A signal from Garrison Point would inform us that a convoy, perhaps thirty, forty or more ships, was due at such and such a time. The small *Keverne*, now in service again, would go off to lie near the Medway Channel ready to deal with ships bound up that river. The three *Danubes* would station themselves a mile or so apart, stern on to the convoy, with engines going slow ahead as the long line of vessels steamed past.

Fine new ships, rusty old ruins, large and small, they pounded past us; sometimes we went ahead to keep up with them, sometimes we dropped back to the next vessel; sometimes we had them on each side of us. It was dangerous work, since the ships, naturally anxious to leave the wartime perils of the open sea and reach safe harbour, came in at full speed, and more than once a collision was narrowly averted.

Seamen are taught to give other ships a wide berth and this close work was a job for old hands. Stenhouse accordingly handled our ship himself in such circumstances and questioned the master of the vessel steaming alongside. Thus: 'Where from?' howled through a megaphone (loud hailers had not yet appeared in our circles) might be followed by an aside to the quartermaster at the wheel: 'Keep her away; this bloody fool's trying to ram us.' And so on. I was usually occupied noting the details of each ship in the deck log and, at the appropriate moment,

Plate 11:

The officers of HMS *Danube III*. (L to r) The author, Stenhouse, Sullivan and Gilroy.

Plate 12:

The former PLA steam yacht at work in the estuary as HMS *St Katharine*.

singing out the next set of flags in the code. As Stenhouse repeated them through the megaphone, the Royal Marine on watch would hoist the flags at our own halliard so that there would be no mistake. I also kept an eye on our list of suspicious ships. Gilroy would deal with stragglers coming up on the wrong side. Sullivan was mainly occupied with our ship's administration.

Question and answer were generally sufficient when dealing with ships in convoy, for our combined knowledge of merchant shipping, usual destinations and routine cargoes would have made it difficult for an enemy to hoodwink us. But if we were not satisfied (especially when we had been tipped off by Naval Intelligence) we would order the ship to the Examination Anchorage where, later, Gilroy and I, accompanied by our two Royal Marines, the party bristling with revolvers and rifles, would search her thoroughly, examine her papers and question the master.

Although big-ship convoys continued to use the Strait of Dover up to the fall of France, we were largely occupied with London's collier fleets from the north-east coast, which were still carrying millions of tons of coal for gas works, power stations and bunkering depots. Part of our duty was to tally the number of ships reported with the number that actually arrived to be sure no enemy vessel was trying to enter port by surreptitiously tailing on to a convoy.

Despite their wartime anonymity, most of the vessels were familiar to me, but out here, marshalled in huge fleets, disciplined by peremptory flag signals, fresh from attacks above, on or below the sea, I found difficulty in relating them to the more colourful ships I had known in peacetime as individual visitors to river wharf or dock quay.

4

Even the estuary weather as I had formerly known it was now different. The heat wave which had accompanied the outbreak of war continued. Ships would appear above the horizon quivering like a mirage. At night, phosphorescence transformed our wake and waterline into shimmering beauty. The stars seemed to be more numerous and more glittering than I had ever seen them before; once, thinking I saw the masthead light of a distant ship, I gave helm orders to avoid a collision with Venus.

Sometimes we saved our coal by drifting with the tide (something I had never seen done in peacetime), and the sun would show the water to be full of small fish, shrimps and plankton. We were encouraged to try and add to our dull rations by setting fishing lines, but we had no luck. Then we discovered that the Gravesend bawleymen and Leigh fishermen, some of whom were still operating, would gladly swap a bit of their catch for a few lumps of coal.

Apart from special operations which sent us out to sea from time to time, our night cruising soon came to an end. The Port was closed at dusk and commercial traffic was confined to daylight hours. Our bunkers were small and fuel had to be conserved, so our four ships were ordered to anchor in strategic positions at night. This and the hot weather were once nearly our undoing.

On that night *Danube III* occupied the innermost anchorage, surrounded by the barrage balloon barges which protected approaches to the convoy roads off Southend Pier. There was no wind and the air was hot and humid. Static electricity built up until suddenly, one after another, the balloons burst into flames and fell, huge masses of burning hydrogen, into the sea where they continued to flare. Two narrowly missed us; a little nearer and we should have been incinerated.

As the hot weather continued into the late autumn we began to get fogs. If visibility deteriorated enough to stop all movement, we would drop anchor and set a frustrated watch peering into the clammy mist and ringing our fog bell. Under such conditions, the Extended Defences Officer would worry about some of the shallow inshore channels (radar, of course, had not yet appeared in the estuary) and once he sent a decrepit steam picket boat out from Sheerness in which I hopefully carried out a patrol between two invisible buoys.

After a few weeks, the defunct *Kite* was replaced by the PLA steam yacht *St Katharine*, now a grey-painted warship armed with a six-pounder gun. The dignified saloon where I had sometimes helped to entertain the Authority's guests was now the seamen's mess deck; its handsome panelling was soon defaced by carved initials, pin-ups and other ornamentation dear to the hearts of *matelots*. The *St Katharine* was never well behaved in a seaway, nevertheless she was to be the first among us to show her teeth when real war came to the estuary.

Our routine was to remain at sea for five or six days, each ship returning, in rotation, to Sheerness for fuel and water. This allowed about twelve hours shore leave. Blacked-out Sheerness offered little relaxation. A bath and meal at the Britannia Hotel left us many hours to fill at a cinema, the pubs or a rather boozy officers' club. Gilroy and I once drifted into the Fountain Hotel, the traditional Navy port of call outside the dockyard gates. In the bar we found a striking tableau. An elderly captain, RN, was teaching a pretty girl to curtsey while balancing a glass of gin on her head. A young sub-lieutenant, RNVR, whom we knew to be the girl's brand-new husband, was standing in the background metaphorically biting his nails. Discipline triumphed and he said nothing.

5

And all through those early days of war, Stenhouse was carrying out his promise to make a practical seaman of me. Sometimes he would send the quartermaster off watch and order me to take the wheel, and I would spend an hour or two steering compass courses to his orders. At other times, I would act as duty signalman and bend on our halliards the coded convoy flag signals. I learnt the flags in the International Code and was soon able to decode the signal letters of inbound ships when they were still hull down on the horizon. Or I would send slow, careful Morse on the Aldis lamp to Garrison Point; and when we were out of range of the lamp, I would signal with the searchlight shutters.

Under the Commander's tuition, I learnt to judge the speed of the ship and the tide, to estimate the force of the wind on the Beaufort Scale and distances in cables.

I had spells in the engine room where our benign old chief taught me the rudiments of the steam engine. (I was glad that Stenhouse's course of instruction did not get to the point of stoking.) The gun's crew took me through the drill of setting fuses, loading, aiming and firing the twelve-pounder, while the Royal Marines taught me to load, aim, fire, dismantle and reassemble the twin Lewis guns. I was given a little rifle and revolver practice, though ammunition was too scarce for much indulgence of this sort.

Stenhouse turned over to me the job of correcting the charts, and many of my watches below were spent with piles of Admiralty Notices to Mariners which were delivered on board when we returned to Sheerness. At other times, I would suffer sore fingers practising the knots and splices taught me by the coxswain.

I always felt flattered when, after a signal had ordered us out on some special mission, say, to the Gunfleet or Sunk Head in the North Sea, Stenhouse asked: 'What's the course, Pilot?' I would set to work on the charts with parallel rulers and dividers and, with more conviction in my voice than I felt, give courses, distances and buoys. And when the buoys turned up at the right time in the right place, I always felt faintly surprised.

Sometimes Stenhouse would yarn about his life; as an apprentice in the half deck where he and his fellows were so hungry that they successfully conspired to steal a sponge cake from the captain's baby; of fights with Melbourne larrikins; of the old Pacific Barbary Coast. But, above all, he spoke about the Antarctic – the silences, the colour, the whales and the penguins, and comradeship against the ice.

Stenhouse was the epitome of patience in his teaching, but at other times he could explode suddenly. Once, when I had been about eight hours on watch and was tired, cold, hungry and wet, he came on the bridge and said commiseratingly: 'A sailor's life's a dog's life, isn't it, Sub?' 'My god, sir, you're right,' was my heartfelt reply. And he blew up. 'What the hell d'you mean by "You're right"?' he roared. 'You ought to be proud to be here, serving your country.'

He was something of a martinet where the seamen were concerned. He barked at them if his orders were not carried out at the double, and gave me hell if he thought I had passed over any slackness on their part. Thinking, he maintained, is the officer's responsibility and all that is needed from seamen is instant obedience. The time of the shellbacks in which he had mastered his profession was closely related to the old dock companies' era of rule by fear.

The Commander must have been reasonably satisfied with his work on me, for when my probationary period was up in the spring of 1940 he sent in the recommendation which brought me my second stripe.

In the process of becoming a seaman, I learnt to appreciate another aspect of the Port of London. Hitherto, I had regarded it as a place of departure, where one looked outwards. Now it was also a haven, a landfall leading to home.

chapter twelve

Mines in the Fairway

1

THANKS TO OUR over-cautious betters ashore, we were robbed of the chance - the best we ever had - to use our guns to deadly effect during the first magnetic-mine laying raid in the estuary.

On the afternoon of 22 November 1939, the Aldis lamp on board *BV10* (the telephone ship) summoned us alongside - literally 'You're wanted on the 'phone'. Stenhouse took the call and hurried back to us.

'Course for the *Barrow Deep* lightship, Pilot,' he demanded. I set to work and we were soon heading out to sea at full speed.

'There's a flap on,' he told us. 'They expect trouble tonight, but they won't say what. We're to tell the lightship to black out and then, on the way back, we're to put out what buoys are still lit.'

Dusk was falling when we hove-to alongside the lightship and passed the order to the master. Then we turned back down the Barrow Deep.

The night was very dark when we went alongside one buoy whose intermittent flashes momentarily lighted up the ship like a stage set and then plunged us into blinding blackness. A Royal Naval Reserve seaman, who had served aboard Trinity House steamers and claimed to know something about the works in light buoys, leapt on to the surging platform while we discretely withdrew to the darkness outside the range of the light and anchored. A boat was lowered and sent away to pick up the seaman who reported that he could not find a way into the buoy's interior and that the glass of the lantern had defied attempts to break it with the heel of his seaboot.

Stenhouse sent him back with a roll of gunny and a hank of codline. The sacking was wound round the glass and secured by the line so that the flashes were now visible only a few feet away. Afterwards, Stenhouse decided to stay at anchor for the night rather than risk the dangers of returning through the blacked-out estuary.

I kept the first watch (eight p.m. to midnight), and the quartermaster, the signalman and I on the bridge, and the gun's crew aft strained ears and eyes, ready to pick up the sound of an engine or sight of the white bow wave of approaching craft. Sure enough, at about five bells the roar of a distant engine broke the silence. I closed up the gun's crew, ordered the signalman to the Lewis guns and sent the quartermaster below to shake the Commander. Then we realised that the sound was not coming from seaward but was approaching from the west; and we relaxed. As Stenhouse climbed the bridge ladder an aeroplane swept past us, so low that we could see the lights on the instrument panel in its cockpit, and vanished out to sea.

'He's come down, looking for the light we've dowsed, to fix his position,' said Stenhouse. 'They ought to have warned us that our planes would be about. If he'd been coming in from the sea, we might have plugged him.'

Then the sound of another engine swelled and a second plane, even lower than the first, flashed past us and disappeared to seaward.

'If any more come,' said Stenhouse, 'you'd better shine the Aldis on our masthead; they can't see us in the dark and that bloke nearly hit us.'

2

About half way back to our home waters the next morning, we realised that something unusual had happened. A large convoy, bound out, was at anchor and some miles of stationary ships stretched ahead of us. We began to thread our way through this mass of shipping when Stenhouse had one of his sudden impulses.

'I think you've finished your apprenticeship, Sub,' he said. 'She's all yours.' And he swept up Gilroy and left the bridge to me. A fast tide was running and the anchored vessels left little room for manoeuvre in the fairway. At speed, we wriggled through the crowd until I was beginning to give helm orders with confidence. Then pride took a purler. I heard a gasp from the quartermaster and turned to see him holding up the wheel as if it were a venomous snake. 'Came off in me hand, sir!' he stuttered.

I rang full astern and we missed a big Dutchman ahead of us with only a few feet to spare. The three blasts on the whistle which I had given ('My engines are going astern') brought Stenhouse up to the bridge at a run. We anchored at once and it took our engineers some two hours to rig the hand-steering wheel. When, finally, we got within lamp range of Garrison Point, we received a bad-tempered signal from the Extended Defences Officer asking where the hell we had been. (Largely as a result of this episode, all our ships were fitted with short-wave radio telephones.)

We found Sheerness like a well-stirred hornets' nest. At about 2200 hours, while we had been anchored alongside the unextinguishable buoy, enemy aircraft had attacked the convoy anchorage, machine gunning ships, the Naval Control centre on Southend Pier and our own harmless old *BV10*. Under cover of the attack, parachute magnetic mines had been dropped. The *St Katharine* had caught one of the minelayers in her searchlight and had been the first ship to open fire on the raiders. We were full of envy, and mourned those two enemy aircraft which would have been sitting ducks for our guns had we only known them to be foes.

The sequel to the raid is well known - how two mines were recovered from the foreshore mud by the heroic work of Ouvry and Lewis and their small party from HMS *Vernon*, so enabling our scientists to produce an antidote. All ships were speedily protected by degaussing or 'wiping' - that is, all ships except those in our humble little flotilla because the dockyards were always too busy with more important work.

Experimental mine-destructor ships appeared in the estuary, and sometimes produced unusual effects: for instance, a marlinspike attached by a lanyard to a seaman's belt rose with each electric pulse to stand out horizontally towards an auxiliary magnet on the forecastle. And in at least one such vessel, the cast-iron tongue of the brass bell forward had to be secured to stop it tolling mournfully in time with the pulses.

Aircraft, wreathed in garlands of magnetic cable, swooped low over the water with the same objective. But the most satisfactory method proved to be the double-L electric sweep, towed by suitably equipped minesweepers.

Night minelaying now became part of our life. To instructions for incoming merchantmen, we would add: 'Keep to the north (or the south) of the channel,' according to where the latest fall of mines had been plotted. On moonlight nights, especially when our furnaces were being fired and sending a plume of smoke across the face of the moon, we felt rather naked. But the raiders made no more machine-gun attacks. Sometimes we managed to loose off at a half-glimpsed plane, merely adding to the brief circus of incendiary rounds, star shell and searchlights without results.

Occasionally a spy plane would fly over during the day - a tiny speck far out of range of our guns. Then one of our signalmen, who before being recalled had worked in an aircraft factory, would squint professionally skyward and, with complete conviction, always give the same verdict: 'Messerschmidt hundred and nine, sir'.

3

If my memory serves me aright, it was about this time that the extraordinarily fine weather went into reverse. Easterly gales followed each other with hardly a pause. Night and day the wind sobbed and screamed in the mast and funnel stays, and flags cracked like pistol shots. Constantly seasick and aching in every bone from the buffeting, I was getting heartily tired of what had truly become a dog's life. Then we had a glorious surprise. All our ships were told to prepare for boiler cleaning in turn, and while in dock a week's leave would be given. When we learnt that *Danube III* would have the first week and that it would coincide with the Christmas holiday, we were prepared to forgive the enemy his mines and the estuary its weather.

But shortly before Christmas the gales changed to blizzards and the worst winter for forty-five years descended on us. Our little vessel would climb laboriously up each wave, squatter down the reverse side and ship a sea which immediately froze on the foredeck, the mast, the bridge, the funnel and the watch on duty. The next sea would wash it all off and freeze in its turn.

We wore as many clothes as the human frame could move in, topping off with a towel round the neck to mop up the overflow. Even so, watchkeeping meant a thorough soaking and near frozen hands and feet. Gallons of scalding tea or chocolate were consumed, and I soon learnt that, given a chance, the wind would scoop out and blow away the contents of a mug as it was raised to the lips.

On the morning of Christmas Eve we returned to harbour with our signal flags proclaiming 'Merry Christmas' - a message not appreciated by our consorts who were remaining on duty among the blizzards. Full of good will, Gilroy and I turned in the ship's confidential books to the office ashore and, as we were leaving, Gilroy wished the Duty Commander a merry Christmas. 'Don't be a bloody fool,' was his reply.

Leave for me was a round of indulgence which was, perhaps, a mistake; when we returned to our refurbished (but not degaussed) ship the continuing Arctic conditions were, by comparison, all the harder to endure. Sometimes Sheerness Harbour during January 1940, was covered with ice, and huge floes wandered in the estuary. The watch below was not too bad, for our little wardroom, with steam heating full on and insulated from the fresh air by folds of blackout material, generated a comforting fug that was fed by pipes and cigarettes.

One morning, we found our windlass without life; the pipe carrying a full head of steam to it had frozen solid. The chief had to disconnect it and thaw it out before we could heave up and get under way. To the westward, we could see a convoy anchored off Southend and all the ships seemed to be on fire. They had all lit bundles of oily waste under their windlasses to thaw out frozen ratchets and cogs.

Our work became more dangerous. A convoy would hurtle out of the murk of a blizzard, giving us little time to manoeuvre into position. More than once we escaped collision only by Stenhouse's fine seamanship; and I saw what he meant about the need for instant obedience from our sailors. During the worst of the gales, the *St Katharine*, not built for such stern tests, had sometimes to make a signal requesting permission to return to harbour. And that left us short-handed for watching the inshore swatchways.

One afternoon, in a screaming easterly gale, a signal sent us chasing after a floating mine, reported to be nearing the convoy anchorage on a flood tide. We found it, a pattern of sinister horns projecting from a mat of weed and barnacles, about a mile downstream of the anchored ships. A few rounds from rifles and machine guns sent it to the bottom without exploding it. 'A pity,' observed the coxswain. 'If it had gone off, it would have officially broken one of the rum jars.'

In the evening we discussed the possibility of such a truant mine bumping into us while we were at anchor. I had the first watch and turned in when Gilroy relieved me at midnight. Just as I was falling asleep, there was a heavy thud against the ship's side by my bunk and I heard something grind its way along the waterline. I was out of my bunk and on the bridge in almost one jump. Gilroy chuckled at my startled appearance and pointed to a large ice floe disappearing astern.

4

All our ships were fed with a constant stream of Naval intelligence on a broad front, whether or not it was of local interest. When we received a signal warning all ships that enemy cruisers had left harbour, Stenhouse said wistfully: 'Wouldn't it be fun if we met them and got in a few lucky shots first.' Gilroy and I looked aft at our ancient gun and said nothing.

That afternoon we saw emerging from a snow squall the turrets and impressive upperworks of a cruiser. 'Here comes one of His Majesty's ships - I hope!' said Gilroy. All was well. After the cruiser had correctly answered our challenge, the squall cleared and we saw that she was the *Ajax* - no less, bound in for refit. Her part in the defeat of the *Graf Spee* the previous December was still fresh in our minds and our crew gave her a spontaneous cheer as she passed. It was a pity she was not bound for the upper tideway which uses massed steam whistles to show its love for a hero.

5

The war and the winter dragged on. Merchantmen were appearing with funnels and upperworks riddled by the increased air attacks on east coast convoys. Most of these ships now had their wheelhouses protected by sandbags or plastic blocks, but some were still without guns. We saw the melancholy sight of a blasted lightship being towed in; the result of this senseless attack was the withdrawal of most light vessels from station.

Stenhouse drew official attention to the inadequacy of our armament against dive bombers. The Admiralty's reply was the issue of a line-throwing pistol which discharged a small parachute. It was hoped that the opening of this in the path of a dive bomber would cause the pilot to turn away. We put much faith in this weapon and kept it in a bridge locker ready for use.

When, much later, the time came to use it, we never saw what happened to the parachute. Our attention was diverted to the unfortunate Marine hopping about and sucking his scorched knuckles. Whether or not he mishandled it, I do not know. Either way, we sadly decided that the estuary would have to be defended with our museum-piece gun.

chapter thirteen

The Storm Breaks

1

SOON AFTER THE Nazi attack on the Low Countries, 10 May 1940, we were on our usual patrol in the Oaze Deep when we received a signal that two Dutch warships were on their way into the Thames. They were believed to be refugees but we were warned to take all precautions on intercepting them in case they were in enemy hands. When we closed them some hours later - a destroyer and a small minelayer - we were prepared for anything.

But Stenhouse, after boarding the destroyer and satisfying himself that they were not Trojan horses, sent them into Sheerness, escorted by the *St Katharine*. He told us of the grief and stunned horror with which the Dutch captain had described the ruthless attack on his country. Other refugees soon followed in a small fleet of Dutch merchantmen, including some schuyts.

The grim realisation that the Nazis had once again seized the initiative would have brought us to despair had not Churchill become Prime Minister. Present-day youth, rebelling against most forms of discipline, can have no conception of what Churchill's iron determination meant to a nation in deadly peril. For the frustrated Services, he was a sudden bright light in the pervading gloom.

On 26 May 1940, Operation Dynamo began. We had a perfect pitch for watching the fleet that went out and its return with part of the Army from Dunkirk. But we were resentful and thwarted spectators. The reply to our protests and pleas to join in was that every available warship had gone to Dunkirk and that our little flotilla was now the sole seaborne defence of the Thames.

It was true that Sheerness Harbour had practically emptied overnight. But it soon refilled with a succession of the most diverse fleets ever to come down river. Several of the rescue craft (especially those which had fitted out in Tilbury Dock Basin) went direct to the jumping-off ports - Ramsgate, Dover, Deal. But many made a temporary stay at Sheerness to take in gear and supplies.

The story of the Dunkirk evacuation has been told so often that I mention here only some of the incidents which focused on the Thames. In the docks and as far up river as Teddington, the Navy, helped by PLA officers and the managers of private boatyards, was working round the clock, taking up yachts, launches, ships' lifeboats, lighters, a sludge hopper, a firefloat and so on.

Nearer at hand, Southend sent its 'bob-a-nob' launches, shrimping bawleys and Leigh cockle boats, yachts and its RNLI lifeboat. Some sixteen Thames spritsail barges made the passage. One of them, the *Ethel Everard*, was beached and abandoned at Dunkirk; she was eventually blown up to prevent her falling into enemy hands. A picture of what was left of her, later published by the Nazis, bore a caption that, roughly translated, said: 'This is how we deal with the British Navy'.

Outward went General Steam Navigation ships from Irongate Wharf, Tyne-Tees Shipping vessels from Free Trade Wharf, F. T. Everard coasters from Greenhithe, and several others. With this mixed fleet went some forty of the refugee Dutch schuyts which had been shepherded into the Port by us. Practically all the big ship tugs of the Thames went, too.

Most of the Thames 'Butterfly' fleet was already on war service; they joined the procession, for they had ample accommodation and were ideal for the evacuation. But for me they were peopled with the ghosts of thousands of Londoners who had sailed in them for a breath of the sea (as well as crab teas and ever-open bars). The only Thames paddler lost was the much-loved *Crested Eagle*.

When they started streaming back crammed with troops, our flotilla was employed to give help and protection where it was needed - a pluck here, sailing directions there, making signals for doctors and ambulances to meet some of the ships. The most self-sufficient of the merchantmen were the Dutch schuyts which had been largely manned by RN crews. Some of them arrived back from the beaches with a young officer squatting tailor-fashion on the wheelhouse roof, reading a chart opened out like a newspaper, and conning his vessel from that well-sighted position.

The *Royal Daffodil* returned with a full load of troops and a fearful list. She had been holed below the starboard waterline by an aerial torpedo which failed to explode until it had gone clean through the ship. Her master, Captain G. Johnson, besides taking other measures, had swung out his port lifeboats and filled them with water, thus giving the vessel sufficient list to raise the worst hole above the sea.

A wave of thankfulness swept over the men who had been saved and over us who saw them come home. For a time, excitement submerged all thought of the defeat which the Army had suffered. The very tides seemed charged with emotion.

Returning to Sheerness for refuelling, we found the road leading to the station lined with hundreds of unshaven, dirty and haggard soldiers, many of them asleep on the pavement. Among them moved housewives, members of the women's voluntary services, nurses and schoolgirls with tea, food, bandages and pity, while troop trains, coaches and ambulances gradually carried away the exhausted men.

There were many stories of enthusiastic volunteer civilian crews. One body of patriots, employed by a Thames-side engineering works, had dropped their tools, hired a coach and, led by their manager, arrived at a coastal depot ready to sail.

Charles Alexander (who had carried out that notable salvage feat in 1930, recorded earlier) was ashore at a managerial desk. When the Admiralty asked for six *Sun* tugs for Dunkirk, he resumed command of *Sun IV* and (still in his City clothes) set off for the beaches towing eleven ships' lifeboats; he made four trips in all. Reading his deck log for those nine days, glimpses behind the understatement and the laconic entries about machine gunning, bombing and shelling, stranding and unavoidable collision in the milling about off the beaches, something of what real seamanship, knowledge of shoal and tide, and generations of inherited contempt for danger contributed to the evacuation. Between them, the six *Sun* tugs lifted a total of 1,420 men, and other London tugs did just as well.

Seeing the cloth caps and city bowlers behind the helms, we were taken by the throat with the drama of it all. 'He that outlives this day and comes safe home, will stand a tip-toe when this day is named,' I quoted. But Stenhouse was more cynical: 'It'll all be forgotten when the war is over.'

On the night of 3 June, it was finished. Where the Thames had formerly led outwards to foreign lands, it was now a frontier, a line behind which the enemy lay in wait.

2

By the time France capitulated on 23 June, the euphoria over Dunkirk had been dissipated and only Churchill's oratory kept up our belief in ultimate victory. So much equipment had been lost that shortage of ammunition became acute. When one of our Marines returned from the shore with three full pans of Lewis gun rounds, we asked no questions. And as he lovingly removed Dunkirk sand from the pans, we gloated as if they were personal treasures.

With the Channel coast of France in enemy hands, convoys through the Strait of Dover practically ceased and most big ships bound for London were routed north-about to join the east coast convoys. Much of the Port's normal trade was being transferred elsewhere, but outward past us came many samples of wartime activity in the river and at the docks - new ships and craft and vessels repaired and refitted.

Air attacks on our convoys increased and sometimes aircraft followed the ships into our waters. Then the yell of 'Where bound?' would be drowned by the roar of our gun. But the enemy attempted only spasmodic high-level bombing on these occasions; he never came within range of our machine guns, and our twelve-pounder was like a pea-shooter trying to kill flies.

Night minelaying increased and new measures were taken to deal with it. Some of the paddle steamers, formerly patronised by day trippers, were equipped with the new still-whispered-about radar and armed with Bofors quick-firing anti-aircraft guns. These were served by Army gunners and the ships were consequently known as the Pongo* Navy. Each night they left Sheerness to lie in wait at the mouth of the estuary.

Minesweepers were now working mainly at night, and, as they passed our ship anchored in one of the approach channels, the watch saw merely a slight thickening of the blackness. When they returned before dawn, we had to be ready with the challenge, given and acknowledged with the pin-point blue light of signalling torches.

Apart from the Pongo Navy lying further out, and the minesweepers,

---

* Navalese for soldier.

we were now the bait for the tiger of invasion. Every night the cruisers and destroyers were collected behind the protective boom while we lay at anchor in the estuary. Our orders, if the enemy came, were to make as much noise as possible so as to warn the ships behind us. In a silence which one could almost feel, we would peer into the darkness; and then a recalcitrant pump far below in the engine room would start to say clearly and loudly: 'Cut his throat, cut his throat', over and over again. Our chief brought all his skill to bear on that pump, but its sinister recitation was never completely silenced.

If the invasion came by day, our orders were to attack at the closest possible range, paying particular attention to tank-landing craft. Stenhouse looked at our ancient gun and said: 'Our best chance would be to ram'. When I pointed out that our low freeboard would leave us wide open to boarding, he thought again. 'We've two rifles, four revolvers and the Lewis guns, but not much ammunition,' he mused. Then he brightened. 'Cutlasses!' he exclaimed. But when he applied to the dockyard, not even these medieval weapons were available for the defence of the Port.

3

The Battle of Britain, as far as the estuary was concerned, began in July 1940, and continued through August with concentrated attacks on neighbouring airfields and sporadic raids on industrial targets such as the Thameshaven Oil Wharves. Over the estuary writhed the vapour trails and into the water plunged friend and foe. I have never seen a reliable estimate of the number of planes which fell into the river during the Battle of Britain. To us, it seemed that the reaches between the Lower Hope and the outer estuary became a crowded graveyard for them.

Bombing of the convoys increased and twice we saved our skins by turning the ship into the fountain thrown up by a bomb, so avoiding its fellows. We huffed and puffed with our twelve-pounder pea shooter and one lucky burst under the tail of a raider caused it to lose height, but it recovered and flew out to sea leaving a trail of greasy smoke.

Although, in old age, I am a man of peace and sympathise with the young who demonstrate against the iniquities of war, I cannot find in my heart any condemnation of the fierce and bitter hatred with which we fought the invaders. We saw what their unprovoked and brutal attacks were doing to our merchant seamen and the little homes in our towns and villages, and so a dead German was for us a good German. It was therefore with much satisfaction that we fished the bodies of dead enemies out of the sea and delivered them to *BV10* for collection by Air Force Intelligence.

We were naturally thankful when we were able to rescue our own airmen alive. Once we despaired as a Spitfire flew straight into the sea and disappeared. We steamed to the spot but there was nothing we could do. Then, after what seemed an unbearably long interval, the pilot of the plane popped up beside us, after going down to the seabed before clearing himself of the wreckage. He had been ditched twice that day and when we landed him there was still time for him to fight again before dusk.

Another pilot swam away from a sinking Spitfire and tried to elude our boat's crew by hiding behind a navigation buoy. Catching up with him, they found that he had been hurtling about the sky so much that he thought he was on the enemy's side of the Channel.

The blitz on London started, as everyone knows, on 7 September 1940, with the Thames providing a perfect flight path for many of the invaders. About a thousand passed overhead, too high for us to interfere with them. But when they (or some of them) returned on that hot and sunny afternoon, they were being harried by our fighters and several were forced down low enough for us to do good work with our machine guns.

One of our Royal Marines was firing the guns on the wheelhouse roof and I was standing beside him, directing his fire. A bomber with one engine aflame floated over us and our guns poured a stream of lead into it. Suddenly the Marine let go the gun butts, clapped a hand to his neck and cried dramatically: 'They've got me, sir!' I was shaken: for us to fire at the enemy was one thing; for him to fire back was another. But then came anticlimax. With the air of a conjuror whose trick has gone wrong, the Marine produced from the open neck of his tunic one of our own hot empty cartridge cases.

Lone bombers above the melee received salvoes from merchantmen in the anchorage (most of whom were now armed) and from the guns of our flotilla. We all made a satisfactory noise but did little if any damage to those high-level aircraft. Our twelve-pounder, traversing to follow the flight of a bomber, carried away the stop that prevented its arc of fire from imperilling our own upper-works, and the blast severed a jackstay, and Stenhouse, standing in the exposed bridge wing, just escaped having his head blown off.

This free-for-all by the ships brought a touchy protest from the Army gunners ashore. As a result, we received a signal that daylight firing on an enemy plane by ships in harbour was prohibited unless the vessels were actually being attacked. A day or two later, while we were in Sheerness awaiting the trot boat to get ashore, the sirens sounded and soon, away to the eastward, we saw an armada of planes that blackened the sky. We waited in vain for our fighters to appear or the guns of the Army to open up.

The senior ship in harbour was a county-class cruiser and she was surrounded by destroyers, fleet minesweepers and a host of minor craft such as our little vessel. The silence in which we watched the approaching enemy was suddenly broken by an order broadcast by loudspeaker from the cruiser: 'All ships open fire!' There was a great roar of gunfire and a storm of shells burst just in front of the oncoming horde. We saw the enemy swerve away over the estuary, let go their bombs into the sea and turn tail. The cheer that went up from the ships marked for me one of the most heartening and dramatic moments of the war.

4

The minelayers now came over nearly every night. We would wait in the silent darkness, guns' crews closed up, the ship completely blacked out. Then the drone of a distant engine would begin and our searchlight, with a dozen others, would pop out like a genie from a bottle. The reception committee, as it was called, behind the boom - cruisers, destroyers, sloops - would open up. The dark waters sliding past were thick with memories of great events, great ships, great men, while around us new traditions were being created in the leaping flashes of gunfire, the tracers climbing up with a deceptively lazy air, and the thunder of the guns. Even our own ancient weapons, alternately lighting up every object on deck and then plunging us into darkness again, were contributing to the Thames story.

Acoustic mines were now being laid along with the earlier magnetic

Plate 13:

Scene in the Port of London on 7 September 1940, when more than a thousand German aircraft attacked the docks and wharves. This was the prelude to fifty-seven consecutive nights of bombing. *Photo: Imperial War Museum.*

Plate 14:

On 25 September 1940, Winston Churchill inspected the damage to docks and wharves. He is here seen embarking at Tower Pier in a PLA launch for the voyage down river. *Photo: Imperial War Museum.*

type, so some of the sweepers battered our eardrums with Kango hammers like any road construction gang. The number of ships sunk or damaged was growing; sometimes we were able to help casualties and sometimes stricken vessels could be beached on one of the estuary shoal patches. The recovery of these ships as well as the raising of sunken vessels was started by the PLA Salvage Service in the summer of 1941 and continued beyond the end of the war (when I had an opportunity of seeing the last stages of this epic programme).

On the morning of 13 October 1940, we steamed off to take station for an expected convoy. At eight a.m. I went up to the bridge to relieve Stenhouse for the forenoon watch, but he was too much interested in a mined and stranded sweeper inshore to go below. We were peering at her through binoculars when I suddenly found myself falling from what seemed a great height in a shower of glass from the wheelhouse windows. I crashed to the bridge deck, entangled with Stenhouse and the coxswain. The tug pitched onto her nose, settled back and then rolled her gunwales under.

We picked ourselves up. Neither Stenhouse nor the coxswain showed signs of damage, although later it was found that both had severely strained backs. Their first task was to rescue the quartermaster lying stunned in the wheelhouse with the huge hand-steering wheel smashed over his head and the wreckage of the steering engine on top of him. I tried to help but I could feel the grating of broken bones in my right leg and I stood, stork-like, on the other, hanging on to the bridge rail.

Aft, the ship was awash as far as the engine-room casing, and a jagged rent ran across the main deck. Both boats were smashed and davits were twisted like wire puzzles; the carley floats had simply disappeared. From the fiddley came flame and smoke. One of our seamen lay unconscious beside a splintered boat. He was picked up by Stenhouse and the coxswain, but as they lowered him to the main deck, the ship slowly rolled over to port.

'Abandon ship!' cried Stenhouse. Then a huge wave swept over us and I was carried down under the halliards and a mast stay.

I am a good swimmer and soon got clear, but two gratings shot up from the sinking ship and took me in the ribs, winding me for a few moments. There came two more explosions which hit us like hammer blows; they were thought to be further samples of the night's drop of mines, set off, perhaps, by our wreckage. Then it was all peaceful and

I found myself floating with the other survivors, all of us clinging to gratings or pieces of the boats. Stenhouse was supporting the still unconscious seaman. The wily coxswain, who had been through many of the big naval actions of the First World War without a scratch, was seated smugly on the ship's big wooden meat safe.

Everyone was calm and very polite. The coxswain offered his meat safe to one of the officers who refused with a 'No, thanks very much, Cox'n'. I asked one of the ratings if he was all right, and I remember the bloody gash which had been his mouth (he had been blown against the galley stove) splitting into a reassuring grin. We were in the water about half an hour, then most of us were picked up by HMS *Janetha IV*, which sent away a boat. The remainder were fished out by our consort, *Danube V*.

The ratings in the engine room, stokehold and mess deck, with one exception, were lost, apparently killed outright. The survivor was a young seaman who had been blown through the mess deck hatchway into the sea. (After hospital treatment, alas, he was discharged paralysed from the waist down.) All on deck had been saved except for our much-respected chief engineer who had been killed on the after deck. Those forward in the officers' quarters had escaped.

Confined together in our little ship for more than a year, each man was a personality, not merely a name, so the loss was all the sharper. Never again would I hear 'Steady it is, sir,' from two of our three quartermasters, or 'Messerschmidt hundred and nine, sir,' from our younger Royal Marine.

Twelve of us had been plucked from the sea and one Royal Marine had saved himself by overstaying special leave when the ship had left Sheerness; thirteen were left out of a total of twenty-two and, for what it is worth, it was the thirteenth day of the thirteenth month of the war.

But we were not safe yet. *Janetha IV* had suffered damage, either from the two later explosions or through hitting our wreckage, and she was slowly sinking as we raced for harbour. She was beached, nearly awash at the entrance, and a launch picked us up and took us to Cornwallis Pier where we were met by doctors and ambulances. There I was given a shot; as I floated into oblivion, I thought, not without some pleasure, that this was the end of my war service to the Port. How wrong I was!

chapter fourteen

Watch on the Middle River

1

THE ROYAL NAVAL Hospital, Chatham, where Dan Sullivan and I shared a room, was still too close to the Thames for our liking. Every night the anti-aircraft guns thundered and we felt that the enemy was now making a dead set at the pair of us. Dan, badly shaken, was allowed out of bed for some hours each day, but I was immobilised in a load of plaster with a broken leg, a fractured heel and some cracked ribs.

We were soon transferred to the wartime Royal Naval wing of a mainly civilian hospital at Dartford, its grounds almost bordering the river. The fifty-seven consecutive nights of raids on London were at their peak and I lay in a state of acute funk, not relieved by the knowledge that near us was a munitions factory and that, from the air, our scattered hutments must look like an industrial complex.

It was not long before Dan was discharged and a medical board found him unfit for further service. He returned to the Thames Nautical Training College and was put in charge of the *Cutty Sark* (then lying at Greenhithe). But, at his age, our disaster had been too much even for his gallant Irish spirit and he died before the end of the war.

My last link with the wartime estuary broke when Stenhouse, after a flying visit to see whether there was any hope of my joining him, left to take up an appointment to the staff of the Commander-in-Chief, East Indies, for salvage duties. Some months later, his ship was mined in the Red Sea and this time, alas, he was among those lost.

The dreary winter seemed endless; my heel refused to mend and nightly the bombers droned overhead. I used to find relaxation in listening to the whistles of ships in the river and visualising the manoeuvres they signified. Other wounded or sick officers came and went until I was the oldest inhabitant of our ward.

At last, in April 1941, I was discharged, fit only for shore duty and with a month's leave. Then I received an appointment to the staff of the Flag Officer in Charge, London, who was housed in the PLA head offices, one floor above where I had worked in peacetime. I was now to have an opportunity to see what was happening in the middle and upper reaches of the tidal river and at the docks.

2

I reported to Flag London on 11 May 1941. There was no warm welcome for the prodigal son; on the previous night, the enemy had made one of his major fire raids. What I saw when I emerged from Mark Lane (now Tower Hill) underground station almost made me wish to be back in the comparative security of the estuary. Along Great Tower Street, Eastcheap and the little lanes running down to the river, the flames raged out of control of the exhausted fire services. Webs of hose crossed the roads and the air was thick with greasy smoke. A warehouse full of foodstuffs beside the PLA offices was on fire and the surrounding streets were rivers of smoking fat.

The PLA building had received its worst damage on the night of the 8 December when the rotunda had been completely destroyed. During this last raid, the top floor of one wing had been demolished.

I found that those members of the PLA staff who had not been evacuated were grim-faced professionals - veterans of the long winter blitz. Many of them had shift-manned the Control Room of the Port Emergency Committee and had co-ordinated the work of safeguarding ships, installations and cargo throughout the Port. Others had served as roof spotters, air raid wardens and Local Defence Volunteers (later Home Guard).

Port operation as I had known it was completely changed. Before the war, lighterage and towage companies, wharfingers and many other Thames-side industries had thrived on competition - and fairly ruthless competition at that, for the tideway has never been a river for soft centres. The nation's wartime needs now demanded co-operation and service of a high order - something which the rather tough operators along the Thames had to learn. That this was so successfully achieved was among the triumphs of the Port Emergency Committee.

With the immense pool of skill and experience at its disposal, it had handled prize ships and contraband cargoes for the Government; dispersed foodstuffs and other essential commodities to 'buffer' depots established and manned by Thamesmen outside London; stored stocks of flour in hatched barges moored in the safer upper reaches of the river; organised and operated a civil defence scheme for the whole Port; and run a river passenger service on behalf of the London Passenger Transport Board. Never before had the Port reacted to a national crisis with such determination.

Later, the Committee also recovered cargo from blitzed dock warehouses and where possible this was treated so as to be usable. For example, out of some 600,000 tons of foodstuffs affected, about 480,000 tons were made fit either for human consumption or for agricultural uses. And about a million tons of rubble from London's ruined buildings were taken away by Thames craft; some 70,000 cubic yards by sailing barges to the east coast to serve as hard core in the construction of new airfields.

3

But this was only part of the story of what, to me, was now an unrecognisable Port of London. The small nucleus of Royal Naval Officers who had moved into the PLA head offices at the outbreak of war had grown into a large and complex command. The name of the base - HMS *Yeoman* - had been inspired by the nearby Tower of London. The Flag Officer in Charge (FOIC) was Rear-Admiral E. C. Boyle, a submarine VC of the First World War.

One of the largest organisations under his command was the Royal Naval Auxiliary Patrol Service, with its own sub-base, HMS *Tower*. It had had its genesis in the River Emergency Service, most of whose craft and men had been taken over by the Navy and employed as general handmaidens of the Port.

Besides two booms laid across the river, the Naval defences included torpedo tubes on some of the wharves in the lower reaches and an extensive controlled minefield. The Navy had also mined certain piers and jetties so as to deny them to any invading forces.

Another section, DEMS (Defensively Equipped Merchant Ships), organised the arming of merchantmen in the Port. By the end of the war, the Fitting Out Gunnery Officer had been responsible for arming some 2,500 ships in the Thames. To train Merchant Service gun crews, a double-decked bus, stripped and painted service grey, was equipped with dummy guns and used as a mobile lecture room.

The Navy degaussed many thousands of ships in London. After being so neutralised, vessels were taken over a testing range either in the Lower Hope or at Tilbury.

The principal Naval Control Service station, responsible for assembling and routing convoys from the Thames, was at Southend, outside the London Naval Command; the famous pier temporarily became HMS *Leigh*, and Naval and Merchant Service seamen, top-secret documents, guns, munitions and casualties passed along the pier railway. More than three thousand convoys were sailed by HMS *Leigh*. For ships using the upper docks and wharves, a Naval Control sub-station was established in the PLA head offices.

At the outbreak of war, the many ship repairers along the tidal Thames came under Admiralty control. Some 1,650 vessels of all types were converted for war and 401 new ships were constructed under supervision of the London Naval Command. Merchantmen salvaged and brought in by the PLA Salvage Service were repaired in the docks and along the river. Altogether, 23,315 vessels and craft of all kinds were handled.

In addition the Flag Officer controlled a vast machine which fed, watered, paid, ammunitioned and served in other ways itinerant Naval vessels using the Port. And throughout the war, a Naval intelligence service was centred in the PLA building.

4

Perhaps because I had been at the receiving end of an enemy mine, it was with these nuisances that I was now to be concerned. Accessible mines, as well as bombs, which had not exploded, were rendered safe by extremely brave and skilful disposal officers on the

Admiral's staff. Mines in the dredged channels were dealt with by minesweepers which daily swept the tidal river from end to end.

But the sweepers could not clear the inshore waters. And since Thames-siders were ordered to take cover during raids, any mines falling close to the riverside wharves, or in the docks, were not seen. This had led to some disasters, notably the loss of the s.s. *Lunula*, a fully laden oil tanker. She had successfully overcome all the hazards of her long passage and had berthed at Thameshaven. Then she had to shift to another jetty and had blown up on an unsuspected magnetic mine; ship, crew, cargo, jetty and tugs had gone up in flames.

It was to help combat this danger that I had been appointed to the staff of Flag London. My new chief was the late Commander Lord Teynham, RN, who became SOMW (Staff Officer Mine Watching) just before I arrived. He was a man of imagination and terrific energy with an unrelenting hatred of red tape, which his upper-crust connections often enabled him to circumvent. As soon as he felt that I had grasped the potentialities of this new mine-watching service, which he had planned in outline, he persuaded the Admiralty to give him a posting afloat.

But their Lordships were not convinced that I was sufficiently experienced to bear the burden of creating this new organisation and so two elderly lieutenant-commanders were successively appointed SOMW. Neither knew the geography of the London Naval Command as well as I did and both were keen to serve in more active spheres. Eventually I was appointed SOMW with full responsibility and in due course given the half stripe of a lieutenant-commander.

5

Our most urgent need was to get a skeleton organisation in being. A young lieutenant RNVR joined me and we spread our net until it included everyone whose business kept him (or her) on Thames-side during air raids; gunsite crews, barrage balloon crews, Home Guard, air-raid wardens, fire watchers, policemen, night watchmen and so on.

Each minewatching post was equipped with a bearing board - a compass rose printed in white on a black disc (for easier visibility in a blackout) round which a pointer, that could be locked by a wing nut, revolved - which we set to true north. The post was given an identification number and plotted on a large-scale chart in our control room.

Thus, on seeing the fall of a mine, the watcher would mark the spot with his pointer and lock it for subsequent checking, telephone us his identification number, the time of the fall and the bearing of the mine from his post. We reckoned that we would receive up to five or six reports of each mine, which would give us an accurate fix of its position.

In the meantime, at the docks and along those reaches where we had been unable to co-opt volunteer night watchers, special posts were built; in due course we received a draft of middle-aged conscript seamen to man them. For the posts upstream of Charing Cross, we recruited Wren ratings. To cover the lower reaches where the river was too wide to be watched adequately from the banks, some sixty superannuated sailing barges were moored at intervals on each side of the dredged channel. The ratings in these isolated craft were administered by the Staff Officer Auxiliary Patrol.

Our next problem was the tube tunnels under the Thames. It was feared that a mine sitting in the mud over any such tunnel might be actuated either acoustically or magnetically by a train, with fearful consequences to the passengers. So we had direct telephone lines laid between the signal boxes and the Naval minewatching posts overlooking the tunnel sites. The tunnel flood gates were closed during a raid and were not opened until our watchers had given a clearance. (A traumatic enough experience for passengers caught below!)

I and my officers went off, one by one, to HMS *Vernon* at Portsmouth to do a course. We learnt to recognise the different types of parachute mines used by the enemy and were shown their individual tricks and booby traps. Useful data were supplied on the splashes made by the fall of a bomb travelling at about five hundred miles an hour, and a parachute mine coming down at only thirty miles per hour.

One of our duties was to destroy mines which could not be swept or rendered safe, and for this we had another course at Portsmouth. I remember the assurance in the voice of the Petty Officer Instructor when he began: 'Modern explosives, gentlemen, are quite safe.' And we saw that he had a finger missing from one hand and two from the other.

The Admiralty presented me with more than a ton of assorted explosives and two portable pulsers. These were enormous generators (towed by lorries) which pulsed through a buoyant cable, laid out from a dinghy. The object was to explode mines in dock basins where recovery was impossible. Only once had we to use the pulser - when a mine was reported in the London Dock; happily it turned out to be a false alarm.

I was like a spider at the centre of this vast web. Two of us kept a night watch in turn, waiting to put our organisation into action. We had a few tip-and-run raids during which the fall of bombs and shells was reported. But the enemy dropped no more mines along the river and we feared that the keenness of our watchers would inevitably decline. We therefore devised fortnightly exercises. And this is where the late Petty Officer A. P. (afterwards Sir Alan) Herbert, MP, came in.

6

A. P. H. had been one of the first to volunteer with his famous yacht *Water Gipsy* for the PLA River Emergency Service, and both had in due course transferred to the Royal Naval Auxiliary Patrol. He and his craft had been seconded for duties with SOMW.

Herbert was tremendously keen always to give that little more than his duties demanded. Once, for instance, when it was rumoured that a parachute mine was visible at low water off Hammersmith, I sent him off in *Water Gipsy* to make a cautious investigation. After some hours, he reported that he had thoroughly prodded the whole area with a boathook and, to his regret, had found nothing. Just as well, I told him, as otherwise there would have been a by-election.

The exercises which we now planned called for a detailed knowledge of the river and good seamanship, both already amply demonstrated by Herbert. Every fortnight we selected a stretch of some seven miles of the tidal river and notified each minewatching post concerned that the exercise would take place on a stated night (with the proviso that if an actual raid clashed with the exercise the latter was automatically cancelled).

Then we plotted some twelve positions in the area where a 'mine' would be 'dropped', leaving to Herbert the job of getting his craft exactly on these spots at exact times, despite the complete blackout, fast tides, barge roads and other hazards.* Herbert has achieved fame as an author and legislator, but I shall always remember him as the man who undertook successfully a task at which many professional watermen would have boggled.

On reaching the first position, Herbert and his crew of two would let a balloon, about five feet in diameter, ascend in the darkness to a height of some seventy feet. The balloon carried in its tail a light which was switched on from the deck when it reached the required height. Still illuminated, it would be rapidly hove down and the light extinguished, and *Water Gipsy* would move on to the next position.

When we had analysed the reports from our watchers we would inform each post how well its minewatchers had done.

The London Minewatching Service was eventually accepted by the Admiralty as a prototype for all ports, at home and abroad, threatened by aerial mining. Its reputation may have been the reason why our parent organisation was never tested by the enemy.

7

During one of the periodical invasion scares, some of us were sent away to take a short 'Shanghai' pistol course, and we had an enjoyable time learning street gunfighting in the best Western traditions. In the mock-up of a village street we were taught to shoot from the hip, to find cover where none seemed available and to hit momentarily-visible life-sized cardboard enemies actuated by springs.

On our return, staff officers were ordered to be armed at all times and we arranged a makeshift range at the south side of the Royal Victoria Dock. Where the Japanese seamen had once played baseball, we passed on what we had learned to our fellows. This included the advice of a petty officer instructor who had said to me when I repeatedly perforated the head of the dummy target: 'You're aiming too high, sir. Hit him in the belly - it's got more stopping power there.'

---

* Some of these are vividly described in Sir Alan Herbert's autobiography 'Independent
  Member' (Methuen, 1950).

Plate 15:

The author with Petty Officer Sir Alan Herbert on board HMS *Water Gipsy.*

Plate 16:

One of the Conundrums, part of the Pipe Line Under the Ocean, moored in the estuary awaiting towage to European shores. Each of these strange craft carried some seventy miles of flexible pipe. By this means, more than a million gallons of oil flowed each day to the Armies of Liberation. *Photo: Imperial War Museum.*

Admiral Boyle's health had been poor for some time, and in 1942 he retired. He was succeeded by Admiral Sir Martin E. Dunbar-Nasmith, a distinguished officer whom, before the war, I had taken round the Port. I was flattered that, without a prompt, he remembered my name (but instant recall of names from the long past seemed to be a faculty enjoyed by all senior officers). He served as Flag Officer in Charge, London, until the end of the war and those who knew him along the Thames still speak of him affectionately as 'The Admiral' and regret his more recent death. He had demonstrated his courgage and resource as commander of the submarine *E11* in the Sea of Marmora (where he, as well as his predecessor, Admiral Boyle, had earned the Victoria Cross) during the First World War, and now he showed his skill in human relations.

An example of his methods concerned the minesweepers repaired and refitted in London. Some of these small ships were still commanded by their former fishing skippers, now holding temporary RNR rank. At the end of the refit, each skipper would be interviewed personally by the Admiral so as to be sure that the work had been done to the skipper's satisfaction. It was a breakdown of the traditional belief in all sea services that those on shore are their enemies. Many old trawlermen left the Thames with a warm feeling that their survival and welfare were matters of concern to high authority.

Of the various captains RN who served the Admiral as Chief of Staff, one of the most likeable and certainly the most notable was Captain Ralph D. Binney. In December 1944, after I had left Flag London, he was killed in a gallant attempt to detain car bandits in a London street. To commemorate his deed, the Binney Medal was instituted and is awarded annually to a member of the public for the bravest act in support of law and order within the London area.

The officer on the Admiral's staff to whom I owed immediate allegiance was Commander Minesweeping. He was a tough old fire-eater with a soft heart. When one of my Wren ratings was reported for being absent from her post of duty, he stormed: 'By god! I'll show her she can't do that sort of thing in the Navy.' But as she was marched before him, the fierceness went out of his face at sight of the frightened little girl who looked as if she ought to be in school uniform. Then he remembered his duty. 'What have you got to say for yourself?' he demanded when the charge had been read out. 'Please, sir, I was drunk,' whispered the girl.

She was let off with a wigging and we just managed to contain our laughter until she was out of the room. In fact, she had had a much better excuse for her absence, but the sentry, an old sweat, had advised her to plead drunkenness as the best method of averting wrath.

The mine and bomb 'busters' were well worthy of tideway traditions. Dedicated men, they spent much of their time in manual toil with their crews, tunnelling after their sinister prizes. One bomb-disposal officer was at the bottom of such a hole at the Surrey Docks when a docker, looking down, asked him how he had come to take up this type of work. 'Influence,' replied our bomb buster without pausing in his labours.

Of the mine busters, one was an Australian old enough to have sons fighting in North Africa. He had lied magnificently about his age to get into the Navy and regarded unexploded mines as some collectors regard butterflies. Therefore, when an unexploded mine just inside his territory was rendered safe by an officer from Chatham, I was not surprised to hear his strident Australian voice on the telephone in the next office bellowing: 'And I'll thank you in future to keep your thieving hands off my mines!'

8

It was some time around 1942 that the Admiralty awakened to the fact that the unorthodox measures taken in London to plot the fall of enemy mines were necessary for other ports, and an elderly commander, RN,* on the staff of the Director of Minesweeping at the Admiralty, was appointed to co-ordinate the work. He had already been twice decorated for mine-recovery work and he was all seaman - from his enormous sheath knife to his grey torpedo beard. His short stature and the beard soon earned him (behind his back) the affectionate name of 'Grumpy', after the Disney character in Snow White.

At Grumpy's instigation, the Admiralty sent me up to Liverpool and Manchester to report on the new mine-watching organisation for the Mersey and the Ship Canal. When I returned and took my report in to

---

*Commander C. E. Hamond, DSO, DSC and Bar, RN.

the Admiral's secretary, a paymaster commander, RN, for onward routing to the Admiralty, he put up his monocle and read it carefully. 'Very clear,' he said approvingly. Then, his service going back to the days when executive officers merely fought ships and did not aspire to coherent writing, he added coldly: 'You ought to be a journalist.'

But Grumpy was pleased and continued to employ me as an ex-officio assistant. Shortly before D-Day, he snatched me away from Flag London to become Base Officer of HMS *Firework*. This 'stone frigate' consisted of two rooms in Vintry House beside the river; there I was in command of two Wrens, a lieutenant and a paymaster sub-lieutenant RNVR, the nucleus of the diving parties - 'human minesweepers' for captured invasion ports. It was the beginning of a new episode which for the only time in my life was eventually to lead me away from the tidal Thames; it has no place in this Port of London record and has been told elsewhere.* But, for some weeks after D-Day, I still had a perch in the Port and was able to see something of the Allied Armada and the beginning of the greatest adventure ever mounted by the old river.

---

*Open the Ports, by J. Grosvenor and L. M. Bates (Kimber, 1956).

chapter fifteen

The Port Strikes Back

1

DURING SERVICE AFLOAT, I had seen the opening stages of war in the lower Thames and the estuary. Now, in Flag London, my duties took me round the docks and along the middle and upper reaches; I was horrified to discover what had been happening while I had been swanning in the estuary.

Passive defence in the Port was an efficient and well-planned extension of existing services. The tentacles of the Emergency Committee spread throughout the docks and riverside wharves to embrace men of all disciplines and grades - dockmasters, engineers, clerks, policemen, watermen and labourers - whose distinctive skills were backed with unsuspected reserves of fortitude.

In the early days of the blitz, the fire brigades had not yet been expanded and merged with the National Fire Service so, although local brigades did what they could, defence of the huge dock warehouses fell largely on their staffs. Fire watching on the roofs was one of the worst jobs. Many of the buildings were literally stuffed with inflammable goods, so that a bomb penetrating the temporary roof defences of sandbags might have created an immediate inferno from which there would have been no escape.

Much colourful propaganda was put out during the war about the indomitable spirit of the Cockney; a lot of it, at least so far as the London docker was concerned, was true. He can easily be led into mischief, as some of the post-war wildcat strikes have shown, but no man, neither dock employer nor Hitler, has been able to coerce him.

This was illustrated by one story told me about a dock labourer doing night duty as fire watcher at the warehouse where he worked by day. A heavy shower of incendiaries fell all round him and he knew that, without help, he could not save the building. He rushed down the dock road just as a fire engine - a rare sight at that time - was passing. He stopped the engine, but the crew had orders to deal with another fire of higher priority and began to drive on. So the watcher boarded the engine and, with blasphemous threats of unspeakable violence, forced the fire crew to deal with *his* warehouse first.

2

At the St Katharine Docks I found something resembling an extinct volcano - blackened walls and piles of rubble surrounding a sinister-looking pool. A mass of burning wax and fat from bombed warehouses had floated on the surface of the water, defying all efforts to extinguish it and flaming for several days. But, as I moved warily through the desolation, the dock staff were already recovering much of the unburnt wax for later refining.

Next door, at the London Dock, the story was much the same; narrow dock roads lined with rubble that had once been fortress-like warehouses. Here I was told about a typically English action by the PLA Police. Amidst the horrors of the early raids they had received a cry for help from outside the dock premises and had promptly sent out a party to help rescue some forty horses trapped in burning stables.

There was even worse desolation at the West India Docks where over a million gallons of rum had gone up in flames at the old Rum Quay. The puncheons had exploded like shells, flinging jets of liquid fire in all directions. And the rivers of blazing spirit flowing into the dock water had created a stiff artificial breeze, making the fire-fighters' task even more hazardous. From some of the massive sugar warehouses had crept slow glutinous streams of fire like lava. When I saw what was left, salvage and clearance were well advanced, but, even so, a walk along the quays and roads was like trying to negotiate a flypaper.

The most horrific blaze - one of the worst in industrial history - had been at the Surrey Docks where some 250 acres stacked with timber had first been fired in no less than forty-two places and had then spread into an unapproachable inferno. Hundreds of firemen (some from as far afield as Bristol and Rugby) played hoses on piles of timber as yet untouched trying to contain the fire. But the water turned to steam and the stacks burst into flame in the searing heat. The timber burned for several days and nights, stoked at intervals by further raids.

At the height of the fire, about half a mile of river frontage was blazing from end to end. A. P. Herbert told me that he had passed downstream in the *Water Gipsy* at the time and that, although he had kept to the far side of the river, which is here about a thousand feet wide, paint on his craft had been blistered.

The story at the Royal Docks was only a little less harrowing – of a deluge of fire and high explosive, of abandoned ships being shifted to safer berths by scratch PLA crews, of parties leaving the hard-hit docks to help ship-repair works outside.

Tilbury Docks had escaped fairly lightly, due, it was said, to the enemy's hopes of using them as a base for an invasion force. The heaviest loss of life here occurred when two PLA tugs were destroyed by a mine at the New Entrance Lock.

Along the river the trail of destruction was much the same. The stories here were of burning oil, ships and craft sunk, Tower Pier destroyed by a direct hit, fleets of barges on fire drifting downstream and then, still blazing, returning with the flood tide. No waterman had reason to complain of night navigation in the blackout during those dreadful times.

Many of the old landmarks had gone. Before the war I had prided myself that I could emerge on deck anywhere in the tidal river and immediately fix my position at a glance. Now, in these battered middle reaches, I was no longer so certain.

3

I found the pattern of port working quite different, with comparatively few goods being discharged at the quays. Instead, much of the cargo arrived at safer west coast ports and was then sent by rail or inland waterway to the Royal and West India Docks for distribution to the markets. For example, all London's meat, bacon and certain other imported foodstuffs continued to pass through these docks.

Some goods were, of course, still sent to London by ship, the convoys continuing to be routed north-about and down the North Sea; but there were now unfamiliar emergency routines. In place of some 150 or more different consignments, the cargo in each ship usually consisted of only two or three blocks of supplies, consigned to the relevant Government Ministries. Such methods were essential in wartime, though undoubtedly they tended to create precedents for post-war austerity.

In the urgent quest for more cargo space, normal ways of casing and baling, as well as stowage in ships' holds, had been replaced by ingenious improvisation. And all the exotic imports - perishable fruits and vegetables, fancy goods, spices and so on, which we had come to regard as regular supplies - had, of course, vanished. Sea carriage had become a stark business of maintaining life and liberty.

As Lend-Lease sent increasing ripples of commercial activity through the Port again, shortages in men and equipment became a problem. Cranes and other handling machinery had been loaned to the services and to less-threatened ports, whither some of London's dockers had also been transferred. Then, too, a number of the younger men had joined the forces. This meant that the average age of the remaining dock workers was fifty as compared with the pre-war thirty, while some I talked to were over seventy. Many port executives were serving with the Royal Engineers; many departmental records had been destroyed in the blitz. As I made my way round docks and wharves, I saw unorthodox methods and short cuts which would have outraged some of the old-time port administrators.

The need to conserve labour was behind the creation, in March 1942, of the National Dock Labour Corporation. This statutory body, predecessor of the post-war National Dock Labour Board, took control of all men registered as port transport workers at most of our harbours. It had powers to move them to ports where they were most needed and, in return, dockers for the first time received a guaranteed minimum wage. It was the beginning of the road which has led to practically all the conditions envisaged by Ernest Bevin.

The war was also responsible for another milestone in dockers' welfare. The blitz had demolished many of the local pubs and cafés where they used to buy cheap meals, and rationing made difficult the snack in the bandanna (such as I had shatttered many years ago). Mobile canteens, operated mainly by gallant ladies of the Women's Legion, supplemented later by temporary cafeterias and kitchens in some of the dock sheds, eventually led to industrial canteens in all London's docks.

4

During the rest of my service with Flag London and the start of my new duties with 'human minesweepers', I saw at first hand what the Port contributed to victory in Europe.

Working alongside and to some extent with executives of the Port Emergency Committee, I was able to appreciate some of its less-publicised activities. Rarely did it have the satisfaction of seeing the end product of such work. In 1942, for instance, at the request of the Admiralty, it provided, through its Lighterage Executive, no less than a thousand dumb barges. I saw these craft in tideway shipyards hurriedly being given armoured protection, engines and movable ramps. They were then sheeted and quietly sent off downstream. Afterwards it was reported that they had manoeuvred as a fleet in the Channel and had so much alarmed the enemy that he had kept divisions on the Continental coasts at a time when his generals were demanding reinforcements in Russia.

Once this country had grasped the fact that a new era of mechanical and scientific warfare had dawned, no other nation surpassed the ingenuity shown by British scientists and technologists in devising new weapons and new methods of supply. The Port of London, with its wealth of varied industry and reservoir of know-how, was a centre for many of these projects.

During September 1941, I saw the first of the Maunsell Forts launched at Gravesend, where four were built for the Admiralty and three for the Army. These towers, equipped with radar and quick-firing guns, were towed to and sunk on strategic shoals in the estuary and proved a successful answer to my old enemies, the mine-laying planes. Soon after the first one had been installed, a passing destroyer made the inevitable signal: 'And how are the little princes in the tower this morning?'

Later at Tilbury Docks came the development of 'Pluto' - the pipeline under the ocean which ultimately supplied 172 million gallons of oil at the rate of a million gallons a day to liberation armies on the Continent. The pipe itself - waterproof, flexible and very strong - was designed by a Thames-side submarine cable company. Within a fortnight of being ordered, a trial length was tested in the river by a Post Office cable ship.

The new piping was delivered to Tilbury Docks where it was welded into continuous lengths of four thousand feet; and projectiles were fired

Plate 17:

One of the Maunsell Forts built along the tideway and then towed out to the estuary to be sunk in strategic positions. Their powerful anti-aircraft batteries helped to diminish the menace of aerial mine-laying in the Port's approaches. *Photo: Imperial War Museum.*

Plate 18:

One of the 'Phoenix, units for the 'Mulberry' invasion harbours being constructed in a London dock. More than five miles of these units were built along the Thames. *Photo: Imperial War Museum.*

by compressed air through each section to ensure that there were no obstructions. Six steel drums - code name 'conundrums' - were constructed at the docks. With a diameter of thirty feet, each had a capacity of seventy miles of pipe and, when loaded, weighed about 1,700 tons. They were launched into the main dock and towed down river by Naval tugs to await D-Day.

One of the most successful training schemes was run in 1943 by the Lighterage Executive of the Emergency Committee. Having already reft the lighterage trade of every man who could be spared, the War Office asked the Thames to train landsmen for service in Inland Waterway Transport craft. Normally, it takes about seven years to make a lighterman and never before had his professional secrets been given to outsiders. But these were exceptional times.

A school was established at the Surrey Docks and it was arranged to give practical instruction in craft on normal work in the river. The Training Officer, the late Charles Braithwaite (of the lighterage company, Braithwaite and Dean) went to an Army depot to select candidates with at least some elementary knowledge of watermanship. He was shaken to find them all raw recruits to the Pioneer Corps, practically all coming from trades that had no connection with navigable water. In despair, he settled for asking each potential student merely whether he could row. Only one man, Charles told me, gave him an encouraging reply: 'Yes,' he had said. 'I've been boatman on the Serpentine for seventeen years.'

In due course, the first class assembled at the school and training began. No less than 1,636 men passed through the school during the thirteen months of its existence, and of these, 1,475 survived the searching individual passing-out test. The top pupil had been a former hotel waiter.

Soon after VE Day some seventy craft manned by the school's former pupils put into the river. In a typical London River gesture, the lighterage trade turned out *en fête,* met and escorted them to an impromptu old boys' reunion in their late quarters at the Surrey Docks.

By the end of 1943, the building of war craft of many types was going on day and night all over the Port, with slipways constructed on waste ground and bombed sites. I saw several tank-landing craft launched at the Royal and Millwall Docks, some with decidedly original messages for Hitler chalked on their plates.

In the same year I saw the start of the 'Mulberry' harbours. More than five miles of their 'phoenix' units were built along the river or in the docks.

These caissons were virtually concrete boxes which were towed to the French coast and sunk on carefully planned sites. To give the least resistance and freedom from yawing under tow, they had 'swim' ends, i.e., sloping inwards, on the traditional lines of Thames dumb barges. They were towed out of their nests partially completed and were fitted out in the wet docks.

The various contractors employed large numbers of Irish labourers. One dock policeman at the Surrey group complained that whenever he asked to see a pass he found it made out for 'Muldoon'. 'All the Muldoons in the world must be working here,' he grumbled, 'and I can't tell one from the other.'

5

At the beginning of 1944 vast quantities of war supplies began to pour into the Port; warehouses and sheds were crammed; higher and wider grew the sheeted dumps. Warships, transports and supply vessels, attended by a host of auxiliary craft, were gathering in the three operational docks – the West India, Royal and Tilbury – and at river moorings. Armed camps were being established for many miles around the Port; suburban streets and village byways were crammed with miscellaneous wheeled transport.

American servicemen and merchant seamen mingled with Thames-siders and accelerated the deterioration of good Cockney started by Hollywood films; and young Cockney jaws moved in unison as they chewed their gum. Either fortuitously or by imaginative intent, the principal American supply base in the Port had been established at a wharf no more than a stone's throw from where Drake had been knighted after the voyage of circumnavigation which did so much to open the New World to English keels.

All the defences of the Port were overhauled in the belief that the enemy might make a desperate attempt to disrupt the forthcoming compaign. The Port Committee extended services and communications to cope with the expected congestion; more dock lighting for night work; emergency dredging; further loading hards; temporary bridges and new dock roads.

No effort in the long history of the Port of London had been comparable to this undertaking. A special 'Western Front' agreement was made with dock labour to ensure continuous and maximum loading despite likely air attacks. A last measure was to wire up the docks and principal wharves so that General Montgomery could explain to thousands of port workers the vital significance of their efforts.

6

Marshalling for D-Day in the Port began on 27 May 1944. Never before had the Thames seen such a fleet of armed merchantmen and ships of war. Between Tilbury and the Isle of Dogs the Port was an immense smooth-running machine which sucked in men, supplies and vehicles and poured them into waiting ships. Some two thousand railway wagons a day streamed into the three dock groups. Transports, landing craft and supply ships came alongside, took in their quota and departed downstream.

In these middle reaches, so long accustomed to the continual coming and going of ships, it was strange to see only vessels outward bound. Port workers became infected with the dramatic implications of the scene and toiled as they had never toiled before. Dockers, too old to fight, bought refreshments for troops at mobile canteens, and more than one bucket of tea went on board a departing ship at the end of a hastily flung line.

Tide by tide, the laden ships gathered at the rendezvous off Southend. There were deep-sea merchantmen, tugs, barges, tankers and several types of landing craft, safeguarded in the anchorage by a flotilla of escort vessels from Sheerness. A further fleet joined them – coasters which had been loaded with supplies and munitions at no less than fifty-four dock berths well before D-Day and which had since been waiting at various moorings.

By the evening of 5 June, the last man had been embarked, the final vehicle had negotiated the ramp, the last ship had anchored at Southend. On the next day, 6 June, the great fleet sailed. In the D-Day sailings from London were 307 vessels carrying some 50,000 Servicemen, nearly 80,000 tons of military stores and about 9,000 vehicles.

7

But London had not finished with the invasion; the Port became the principal supply base for the British Army as it fought its way up Western Europe. There was no respite for the Port Emergency Committee in organising the flow of supplies, the quick turn-round of ships and the provision of fuel barges for instant bunkering; work soon to be hampered by flying bombs.

It was now that I had to leave the Thames for the last year of the war – to keep house for 'human minesweepers'. When the rocket bombs began to fall, I became partly resigned to my stay-at-home role in the invasion. My lasting regret was in not being back on Thames-side for VE Day to hear the traditional massed blowing of ships' whistles.

It was estimated at the end of the war that approximately 15,000 high explosive bombs, 350 parachute mines, 550 flying bombs and 240 rockets, as well as countless incendiary bombs, had been dropped on London's dockland during some 1,400 raids. Damage to PLA premises and equipment was eventually assessed at a pre-war valuation of £13½ million, while the private wharves and other independent operators accounted for another large sum.

But the greatest damage to the Port had occurred in the approaches where mine, bomb and torpedo drove much of its normal trade to safer ports. On 25 June 1941, Winston Churchill, in secret session, told the House of Commons that London's overseas trade had been reduced to a quarter of its usual volume. Only 106 million net register tonnage of shipping used the Port between September 1939 and May 1945 as compared with a yearly total of some 60 million tons.

SECTION THREE

TOWARDS THE FUTURE

chapter sixteen

Post-War Frustration

1

DURING THE AUTUMN of 1945 I returned to the Port and to civilian life; once again I was involved in the commercial world of London River. The victory euphoria had already gone, and disillusionment had set in as far as the nation was concerned. But while I went round the docks and along the Thames, I was amazed at the resilience and the absence of despair.

God knows there was little enough to be hopeful about: the skyline still showed many gaps, torn brickwork and riven concrete; about half the Port's warehouses had vanished; much equipment, commandeered by the Government, would never come back, and what remained on the quays needed much more than a mere lick of paint; shoaling in the river fairway cried out for maintenance dredging, largely halted during the war; wrecks still littered the estuary channels; many traditional trades showed no signs of ever returning to the Thames.

But such was our blind faith in the Port's significance to the country that we never doubted for a moment that the great liners, the mountains of cargo, the torrent of trade would soon return. Oh yes, we agreed, there would certainly be changes, but they would be merely changes in detail. In principle, it would all be much as it was before. My friends along the river gradually returned, more sophisticated, more demanding, but everyone expected them to settle down to civilian life and accept the proper station that God had been pleased to give them.

For the second time in the life of the PLA, war had put the Port back to square one. But the Authority which emerged from the struggle had changed nearly as much as the Port. Many of its long-standing convictions had been rejected and its young men were bringing back fresh ideas after service in foreign ports. More pragmatic and less orthodox methods were valuable spin-offs from the war. The magic carpet of the post-war years was to be the forklift truck and its kinsman, the mobile crane, both used by the forces to such good effect while stocking up for D-Day. These new cargo-handling tools were envisaged as operating not only on the quays but in the sheds and warehouses, too, where they could stow cargo more economically. And this meant that sheds and warehouses would have to be designed with wider doors and with as few stanchions and pillars as possible to allow the trucks to manoeuvre freely.

If there was one thing learnt by this nation in the war, it was the need for expert planning; the national penchant for amateurism had no place in the post-war atmosphere, and we at last abandoned the 'It'll come right on the night' philosophy. Now the future of the Port was to be planned to the last foreseeable detail. The height, length, breadth and shape of all new building, as well as relevant dock layouts, the often-conflicting needs of road and rail transport, the fluctuations in seasonal trades and the sizes of ships – these and other factors influencing efficient operations were exhaustively studied before a T-square was put to paper.

All such planning was based on the assumption that there would be little alteration in the basic methods of sea carriage – that the process of handling individual packages to and from ships' holds, traditional at least since the days of the Phoenicians, would continue. In other words, the container revolution was still well below the commercial horizon.

One hint of changes to come was the post-war use at Tilbury Docks of a ramp which had been constructed for D-Day loadings. It was now used by former tank-landing ships, adapted to suit conventional vehicles. These vessels opened their jaws to receive or disgorge laden lorries bound to or from the Continent. Out of this sword turned ploughshare have grown the important and thriving roll-on, roll-off fleets of today.

But the Government soon disposed of any post-war optimism by decreeing that reconstruction of British ports should be at the end of the queue for labour and materials, both desperately scarce, and by refusing to sanction the capital expenditure. A farcical situation developed. Ships were beginning to return to normal trading, and shipowners and manufacturers were being exhorted by the Government to concentrate on exports. All three demanded a quicker turn-round of ships in port. British ports, however, were still refused the means.

Meanwhile many of London's Continental rivals, ports which had in the main been completely devastated, were able to rebuild from the ground up (aided by American capital) on the most modern lines. For

London and other British ports began a grim period of mending and making do, as the current saying went in those days. Large-scale reconstruction and re-equipment did not begin along the Thames until nearly 1950.

2

In 1947 the National Dock Labour Board superseded the wartime National Dock Labour Corporation and set the seal of permanency on Ernest Bevin's great reform. The docker now received attendance money and a guaranteed weekly wage whether or not work was available, and holidays with pay. Proper 'call-on' accommodation replaced the old 'stands', excellent medical centres and ·other welfare amenities were provided. A chain of industrial canteens for lunch, and mobile canteens for the morning and afternoon breaks, were claimed to be among the best in the world. Pay arrangements were centralised so that men no longer had to call for their money at perhaps three different offices. And throughout the Port the new cargo-handling machinery was taking much of the physical effort out of dock work.

But the situation was still not as Utopian as Bevin would have wished. Although many of the old evils had been removed, the work was still basically casual - when there were ships to be dealt with, the men earned good money; when there were no ships, they received only attendance money. It has more than once been said that casual labour produced casual attitudes to the work. Moreover, the National Dock Labour Board, employer of all London dock labour, had its head offices on the Albert Embankment, far from the docks, and, as Lord Waverley* observed: 'There is a blurring of the old employer-worker relationship so that the majority of the men do not know who really is their employer.'

The general post-war disillusionment was eventually reflected in the attitude of port workers. In military terms, we had 'won' the war (although few for long sustained that belief) but there seemed no end to

---

* As Sir John Anderson, appointed PLA Chairman in 1946. Created Viscount 1952 and continued as Chairman until his death in 1958.

the shortages, the rationing and the dreary makeshifts. During the war, there had been magnificent co-operation between worker and employer. Inspired by the enemy at the gate and Churchill's oratory, they had together fire-watched their places of work, mine-watched the tidal river and cheerfully put up with the nightly battering, the poor food and the generally sordid conditions. And some of the Port's more starry-eyed operators had looked forward hopefully to a better post-war industrial atmosphere. But soon after VE Day, the old suspicions and mistrust began to emerge and the comradeship forged during the war faded. Once more it was 'them' and 'us'.

Despite the almost complete reversal of their former role as the waifs of industry to the status of best-paid unskilled labour in the country, militancy continued to grow among the men, especially the younger dockers back from the war. Old jealousies between dockers in the 'pool' (from which men were allocated to employers as required) and the 'perms' (men permanently on the books of a single employer) were intensified, and more than once during industrial disputes there were reports of back-street windows being broken and intimidation of wives of 'perms', and even of their children being beaten up at school.

Conditions of Thames lightermen, too, now bore no relationship to the old days. Like the dockers, they came under the National Dock Labour Board and received pay and enjoyed amenities such as old-time lightermen would have found hard to credit. Now usually voyaging behind craft tugs instead of working the tides over long distances, they were still tough and skilful. But the old professional pride and local loyalties were beginning to die and few showed the independent spirit of their forbears.

Just as after the First World War, khaki became for a time the common working rig of dockers back in civilian life. But the varied British and foreign uniforms which had lent so much colour to dockland during the war soon faded from the memory of the public. Not that the public was ever very knowledgeable about such things. I well remember the rage of one of my senior officers on the Admiral's staff when he was taken for a commissionaire by an imperious lady at a London store.

One of the PLA Harbourmasters now told me of a somewhat similar experience. These men wear a slight variation of Merchant Navy officers' uniform. Going off duty late one night, he sat in a bus beside another passenger who gratuitously informed him that it was very cold and that his daily work took him out into this inclement weather. Whereupon the PLA man mentioned mildly that he, too, found his working conditions a bit tough. The passenger looked respectfully at his uniform and replied: 'Yes. I dunno how you chaps stand the way the wind sweeps round them platforms every time a train comes in.'

3

I was among the first to be released by the Navy after VE Day and I now made cautious sallies along the river and through the docks. I knew that much of what I loved had survived, but that some of the details had altered, and this process was continuous.

Many of the ships I had known had been sunk. Some were still trooping or in other Government services, operating from unfamiliar ports. In all docks and many riverside berths were the wartime mass-produced ships ('built by the mile and cut off by the yard') - the *Liberty, Empire, Fort, Sam* and *Victory* types. Although the Port was conscious of the tremendous wartime debt owed to these vessels, no one was very sorry when their somewhat ungainly forms and often unlovely names were slowly replaced after the war. During the 1950s, I was lucky in being invited to join a number of shake-down cruises of new ships. At first I was critical of their sometimes peculiar-looking funnels, rather as if the ships were wearing carnival hats, but I soon learnt to appreciate the naval architects' attempts to reduce the nuisance of funnel smuts.

4

In the late 1940s, ships were still bringing complete cargoes consigned to various Government ministries. And such was our post-war plight that official and private consignments of gift food were passing through the Port from commonwealth and foreign countries up to about 1950.

It was in January 1946, that the s.s. *Jamaica Producer*, the first banana ship to discharge in London after the ban on such imports was lifted, berthed in the West India Docks. The first stem of bananas to be landed was ceremonially greeted by the Mayor of Poplar.

In March 1948, the Lykes Lines *Bluefield Victory* brought a large consignment of rice, a personal gift to the people of Britain by the owners of an American rice farm. Back to the States in the same ship went a gift from the RAF - the beaver boards embellished with pin-ups and graffiti which had once graced the wartime camp of the USA's 401st Bomber Squadron while stationed in this country. Remembering its camp with nostalgia, the Squadron had asked if it could have the boards.

Another reflection of the war was the arrival of large consignments of raw rubber from Malaya. The bales had been confiscated and stock-piled by the invading Japanese, and in the course of time many of the rubber sheets had fused. In some cases, the metal retaining bands had disappeared into shapeless lumps of rubber. To avoid damaging manu-facturers' machinery, PLA staff were trained to use mine detectors and so discover any metal in the bales.

An undertone of the D-Day invasion was a small sandstone boulder which I saw lying beside the drydock at Tilbury. One of the invasion craft had holed herself so badly when she touched down on a French beach that she had to be dry-docked at Tilbury on her return. The boulder was removed from her damaged bottom plates - the first piece of liberated France to come to Britain.

5

Perhaps the greatest change in the post-war river scene was the eclipse of the spritsail barge. Before the First World War, some two to three thousand of these craft carried freight up and down the tideway or between east and south coast ports, while several regularly traded to and from the Continent. Between the wars, the number under sail without engines was a little less than a thousand; their red-brown sails were still very much a part of the river pattern. At the end of the last war, the total fleet of spritsail barges without auxiliary power had dropped to less than thirty. And within a few years, the commercial sailing barge had virtually disappeared.

Like some other venerable trades in difficulties, the main reason for the decline of these craft was manpower, although a contributory factor was that during the war the internal combustion engine had fostered new conceptions of time-saving. The old race of sailormen was dying out and few recruits were attracted to a profession which involved cold and wet conditions, irregular pay and long hours, much hard physical work and some personal hazard. No one could blame the barge sailors for quitting their dangerous tide-washed decks, but when the spritsails finally vanished (with the exception of the few barges used as training vessels or sailing yachts) the Port lost something of value.

While the frustrated PLA contemplated the damaged or destroyed quays and warehouses about which nothing could as yet be done, it was able to tackle one vital job – clearance of the remaining estuary wrecks.

In the normal way, the Authority had adequate plant and highly experienced men ready to keep all channels in the Port clear of 'wreck and obstruction'. The plant consisted mainly of powerful wreck-raising ships and a number of salvage lighters which could be secured on either side of a wreck at low water and which, with the wreck cradled between them, would be lifted by the rising tide. Lighters and wreck would then be towed to a suitable beaching place. In command of this fleet was the PLA Mooring and Wreck Raising Officer.

At the outbreak of war in 1939, the PLA, its salvage fleet enlarged by extra Admiralty vessels, had undertaken to keep the estuary approach channels, normally outside the then Port limit, clear of wrecks. The wartime record of this salvage team was magnificent. Despite mines beneath the water and bombers in the air above, ship after ship was raised from the fairway or the enclosed docks and towed to repair yards where they were made ready to re-enter the vital wartime trade of sea carriage. More than once the salvage vessels were pinned by their wires to a submerged wreck when they were attacked by enemy aircraft. And always there was the fear that if an invaluable salvage craft herself fell victim to a mine not only would her expert crew be in hazard but she would be irreplaceable under many months.

There was a lighter side, too. The Salvage Officer once incurred the displeasure of the Ministry of Agriculture by pushing into the river a cow and her calf which swam ashore from a Greek ship sinking after a collision off Woolwich. The Ministry acidly referred to the episode as 'the unauthorised landing of foreign cattle'.

In 1946 I went down to the estuary to watch some of the post-war salvage work. About half a mile south-west of Number Five Yantlet Buoy I found the twin salvage vessels, *Kinloss* and *Help*, anchored over the bones of the s.s. *Belvedere*, mined and sunk in 1940. She had been loaded with a complete cargo of cement, long since turned to solid concrete and embedded by some nine thousand tides deep in the ooze.

A red flag beside the routine green wreck flag and its two green shapes told passing ships that divers were down and requested a reduction of speed to avoid creating a dangerous swell. The wartime advances in self-contained diving gear had not yet reached commercial operations and so the divers here were all using the then conventional pumped air supplies. Amidst the tangle of torn plating and smashed upperworks, they were trying to secure wires on which the salvage ships could heave.

The visibility of these divers on a completely cloud-free day would be about three inches; for the greater part of the time it was nil, and the divers could work only by touch. Moreover, the strength of the tide meant that they could go down for only about one hour on each side of slack water.

To me, with some little knowledge of wartime clearance diving, this operation amidst badly damaged wreckage seemed to pose considerable dangers. But the divers took it all as a matter of course. One of them told me that his most frightening experience was when he and his gear were transported from one dock to another. The van in which he made the journey was driven by a youth who exceeded the speed limit and broke most other traffic laws during the run. When, thankfully, the diver reached his destination, the youth peeled the fag-end from his lip and said in awed tones: 'I reckon your job's pretty dangerous, ain't it, Guv'nor?'

On the day of my visit, the forlorn topmast of the s.s. *Araby*, mined and sunk in 1940, could be seen seaward; nearby was the mast of the s.s. *Beneficent*. Near the Nore Sand buoy, salvage lighters were struggling to lift the collier *Pinewood*, mined in 1941. Over towards the Medway Channel were the upperworks of the sunken *Richard Montgomery*, wrecked during the D-Day sailings. She was fully laden with high explosive and was such a hot potato that the PLA salvage team was not allowed to touch her. More than thirty years later, her rusting carcase (and her cargo) remains, a potential menace to passing ships.

This clearance of wartime wrecks was not finished until 1950. During the preceding years, the PLA salvage team had raised thirty-five ships and 598 small craft (barges, tugs and so on); dispersed eight (including my old wartime ship HMS *Danube III*); and helped to save a further forty-six.

Near the end of the 1940s, the foaming beer-coloured liquid that sluiced past the piers and barge roads not only stank as never before, but was also rejecting further attempts to deepen and in some areas even to maintain the existing depths of the dredged channel. It was known that pollution and siltation were linked, largely due to the same causes - the overburden of sewage effluent, (500 million gallons daily) mostly only partly treated, which robbed the water of its oxygen and caused it to drop its load of silt instead of carrying it down to the sea; the dumping of industrial waste; the amount of fresh water taken from the non-tidal river for the supply of London; and the raising of river temperatures by the discharge of huge volumes of hot water from the growing number of riverside electricity generating stations.

This situation illustrates some of the dilemmas that confront long-term planning. The dredged channel was responsible for the Port's increased trade and so a major reason for London's expansion and rise in population. In turn this demanded more power, more water supplies, more sewage works - all contributing to pollution and siltation of the river at a time when still further development of the Thames was essential.

Lord Waverley had a considerable scientific background, and he set up a special committee and brought in the Government Department of Scientific and Industrial Research. A full investigation of these allied problems of pollution and siltation was begun. In the course of the inquiry, minute quantities of radio-active mud were put into the river so as to trace the movement of silt.

A large-scale working model of the river bed - 400ft long - and its tides (generated electronically), complete with 'model' silt, was built in a shed at the Royal Victoria Dock. To watch the tide ebbing and flowing through the channel of the model was fascinating. The full twelve and a half hour tide cycle was condensed to thirteen and a half minutes. When my ship was running out of Sheerness during the war, we had sometimes encountered a powerful eddy at certain states of the tide where the waters of the river Medway met those of the Thames. I now saw this eddy faithfully reproduced in miniature on the model.

In 1957 the committee's report (backed by a later report on effluent by a Government committee under Professor A. J. S. Pippard) resulted in several important measures. The LCC (later GLC) pressed on with improvements at their sewage outfalls at Beckton and Crossness in the middle tideway. Full co-operation was given by Thames-side undertakings to stop polluting discharges. And the PLA changed its long-established method of dumping silt brought up by its dredgers, silt which the inquiry proved soon returned to the fairway from the dumping ground in the Black Deep. Henceforth it was pumped ashore for marsh reclamation.

This more scientific appreciation of the regime of the river has substantially reduced costly maintenance dredging and improved the depths of vital channels; equally important, the tidal Thames has become a cleaner, healthier waterway. In the early 1950s I had seen large numbers of dace breaking the surface of the water at Teddington, and I had mentioned this to the PLA River Purification Officer, adding that I was tempted to take a fly rod there if the fish continued to rise so freely. 'They're not feeding,' the officer had replied sourly. 'They're suffocating and are trying to get more oxygen.' In the middle reaches where the water was particularly poisonous, it is doubtful whether at that time a fish of any kind could have long survived.

Today wildfowl are returning to the river in ever greater numbers, and there are now some ninety different types of fish, both fresh and saltwater types, in the tideway. Rainbow trout have been caught in enclosed docks, despite the traffic of ships and barges. The latest type of fish to be caught is a seahorse.

The PLA has provided a trophy to be awarded annually for the largest salmon caught by rod in the river between Teddington and the estuary. Salmon parr have been put into the upper river and we may yet see this fish return to the Thames in such quantities as when London apprentices demanded that they should not be fed on this fish more than a stipulated number of days each week. But a cautionary note by an expert points out that the Thames, as an industrialised lowland river, is unlikely ever to become a great salmon water.

On 1 April 1974, statutory control of river purification along the tideway passed from the PLA to the newly constituted Thames Water Authority who, with the aid of the most modern scientific equipment, carry on the good work.

chapter seventeen

The Port Stirs

1

EARLY IN 1953 death and destruction spread along Thames-side when a violent northerly gale drove the flood tide (higher than ever before recorded) over and through sea and river walls along much of the east coast and into the Thames, upstream as far as Putney. The heaviest casualties were on Canvey Island in the lower tideway. Here the Dutch-built river walls which had withstood earlier attacks of wind and water for more than three centuries (when defences elsewhere had failed) were breached by this great storm surge, causing utter desolation. At Tilbury, Gravesend, Foulness, Sheerness and the Isle of Grain, too, many inhabitants had to be evacuated and some riverside works temporarily closed.

National appeals and a distress fund opened by London's Lord Mayor brought a generous response in the way of comforts for the homeless. The collection and distribution of these were centred at the PLA's warehouses in Cutler Street (that old legacy from the Honourable East India Company).

There followed a Government Committee of Inquiry, headed by Lord Waverley. Its principal finding - that a moveable tidal barrier should be constructed - started a long public debate about design, costs, responsibility and a suitable site. Despite many warnings of more serious flooding in London, it was only some twenty years after the disaster that work began on a site in Woolwich Reach, where sections of the moveable barrier are now being constructed.*

---

* Due to be completed by 1981.

2

By 1950 most of the post-war Government restrictions on reconstruction work had been lifted, giving the PLA its chance to move towards the future. There followed the scrapping of certain traditional methods of handling cargo manually, such as the timber trade at the Surrey Commercial Docks where specially adapted mobile cranes were able to work in new, more spacious storage sheds as well as on the quays. Timber porters began to lose the callouses on neck and shoulder, formerly the badge of their trade.

Of wider significance was a successful experiment at the West India Docks where the mechanisation of export cargo was eventually to provide a prototype for similar work elsewhere in the Port. Hitherto, crates and cases for export had been handled manually from arrival at the docks to the loading of the ship. Now, by a combination of mobile cranes, fork-lift trucks and pallets (platforms raised a few inches off the ground, allowing the forks of fork-lift trucks to be thrust beneath them) productivity, that recurrent theme, was increased by some twenty per cent. But, although PLA, shipowners, exporters and even dock workers were satisfied with the results, there was the inevitable snag – at least for English-language purists – the insidious growth of another jargon, mostly hideous in itself but undeniably convenient as a means of communication. The new omnibus word was 'palletisation' but other lesser bastard words include 'packaging', etc. Some horrors of our own time include 'containerisation' and 'jumboise'. So we pay dearly for material progress.

3

Another change along the tidal river was now accelerating – the replacement of steam by diesel, particularly in tugs. For some years the diesel engine had been queening it on the high seas while steam tugs, especially for ship towage, had still persisted in the Port.

Most London River tugmen were notoriously averse to change. Even after the famous marine tug-of-war between HMS *Rattler* and *Alecto* had proved the superiority of the screw-driven vessel over the paddler, the paddle tug had survived in general use for many years. Some tugmen

Plate 19:

Before the Thames Water Authority took over conservation of the tidal Thames in 1974, the PLA River Purification Officer took samples of industrial discharges for analysis. *Photo: Post of London Authority.*

Plate 20:

Artist's impression of the Thames Tidal Barrier in the western half of Woolwich Reach, due to be completed in 1981. *Photo: Greater London Council.*

long asserted (as, in fact, diehards still do) the advantages of the more manoeuvrable paddler in confined waters.

A later breed of tugman boasted the reliability of the steam engine over the internal combustion engine and managed to delay the inevitable introduction of the diesel tug. Today, modern, stream-lined, squat-funnelled diesel tugs, breathing power in every line, are part of the river scene. They can be run with a smaller crew, say six or seven, than a steam tug where the firemen might bring the total crew up to about nine; running and maintenance costs of the diesel craft are lower; the engine can be operated from the bridge, thus ensuring instant response; no time is wasted in raising steam (an important feature of salvage work); and bunkering takes less time.

For its docks, the PLA ordered a fleet of twin-screw diesel tugs to replace its worn-out steam-driven craft. These, in turn, were later replaced by tugs equipped with Voith-Schneider propulsion whose rapid manoeuvrability outclasses even the old paddlers.

Independent tug owners, too, now found themselves afloat in a new type of commercial world. The days when a towage contractor went round the City every Friday afternoon with a bag of sovereigns, paying his debts and collecting his dues, had long since gone. But there had persisted an uncompromising and direct approach to the business of towage which was strangely old-fashioned. Perhaps it had been necessary in dealing with the conservative and sometimes eccentric People of the River.

One incident which illustrated how difficult it was to handle Gravesend tugmen in the old days was told me by the late Charles Etheredge. During the First World War, he had accepted an Admiralty contract to bring to this country some forty big Rhine barges lying at Rotterdam. Mines and enemy submarines made the proposition unattractive to his men, and he promised special bonuses when the job was done.

The first tug made the crossing safely and began the return with her tow. On passage, she paused to help two destroyers damaged and foul of a minefield in the North Sea. Ordered to take his craft out of danger, the tug skipper told the senior naval officer to mind his own lurid business and carried on helping the two warships, whether they liked it or not. For this he later received an official reprimand together with an award for gallantry.

When all the barges had been brought over, the tug crews gathered in Etheredge's office to receive their pay and bonuses. But, heroes all, they could not agree among themselves over the shares and ended up with a bloody free-for-all round the office table and desks.

Now, after the Second World War, London tug companies, some of them with even more recent memories like those of Charles Etheredge, found themseles pushed into modern times. With mergers,* costing accountancy, work study, etc., they are now as efficient and streamlined as their elegant tugs.

4

I have already mentioned that before the war London's largest import was 13 million tons of coal annually. Some forty years later the quantity has dropped to less than 3 million and oil has usurped first place with 21 million tons, representing about fifty per cent of the Port's imports.

By comparison, the collier is now almost a rare visitor to the Thames, but the great fleets which formerly filled the fairway on every tide did not decline without putting up a fight. As late as the 1950s I visited one of Hudson's new colliers which had arrived at Dagenham on her maiden voyage. The old Geordie skipper almost purred as he showed me his day cabin, the companionway to the bridge and other comforts. Only those who had known the stark discomfort and exposure of the old-time colliers, half submerged in the winter North Sea, could have understood his pleasure. 'Proper carpet-slippers job now, mister,' he chuckled. Then he added that his own home was not so grand as this new accommodation and that his wife was jealous.

The modern trade in imported oil virtually began in this country in 1862 when the 224-ton brig, *Elizabeth Watts*, arrived in the Thames from Philadelphia with some 1,200 wooden barrels of oil. The cargo was discharged in the Royal Victoria Dock, and no doubt the master breathed a sigh of relief when the last barrel had gone. With so inflammable a cargo, it was predictable that the *Elizabeth Watts* would have difficulty

---

* The principal Thames ship-towage companies merged as London Tugs Ltd., and are now part of the Alexandra Towing Company Ltd.

in finding a crew. Eventually she was manned largely by a draft of drunks shanghaied on the eve of sailing.

From this epic voyage grew a trade which made London one of the largest oil ports in the world. For many years the main storage centre was at the independent London and Thames Haven Oil Wharves in Sea Reach. But from 1950 onwards the chain of Thames-side oil refineries began to take shape. Hitherto, most of the imported oil had been refined at or near the oil fields and only the end products shipped. New reasons for a reversal of that policy included the recurrent balance-of-payments crises; an upsurge of indiscriminate nationalism in some of the oil-producing countries; an ever-growing appetite in the western world for oil fuels; and, in particular, technical advances in the design and construction of tankers which made the sea carriage of crude oil economically attractive. (During the ten years ending 1966, the cost of carrying crude oil by sea was almost halved.)

Accordingly, from the late 1950s onwards, the growing size of oil tankers has presented most of the world's port authorities with the problem of creating deeper channels such as the PLA is providing. Tankers of over 200,000 d.w.t. have used the Port in recent years; large enough to have carried the *Elizabeth Watts* as an insignificant piece of deck cargo.

Refineries, products of modern technology, have not suffered from the traditional labour problems and inherited customs which have dogged older Thames-side industries and are to some degree a source of envy to progressive port operators fighting a rearguard action against incipient Luddites.

5

But this post-war period was not only notable for pragmatic planning; the tideway still produced those inconsequential incidents that remain its principal charm.

Standing one day on Westminster Pier, for instance, I saw a public launch discharge her load of sightseers. In the stern was a wild duck brooding a clutch of eggs on a nest made of rope. From time to time, I was told, the mother duck went off foraging, sometimes having to chase the launch upstream or down to resume her maternal duties. I was assured that she would not be disturbed and I believe the eggs were safely hatched.

So I knew that the river could be trusted, at least as yet, not to go whoring after too many new-fangled ways; that old customs, held in abeyance during the war, had not been forgotten.

Swan Upping was resumed. The swans on the river are owned, according to custom, either by the Crown or by the Dyers' and Vintners' Companies. In July, watermen in traditional garb and flying pennants go afloat to capture and mark new cygnets. Among their memories was the pair of swans which regarded themselves as river craft and took it as a matter of course that the dockmaster would lock them with the barges into or out of the Surrey Docks.

Thames Watermen's annual race for Doggett's Coat and Badge came back. Not only prowess at sculling is necessary but also an intimate knowledge of the strength and set of tidal currents. Founded by a seventeenth-century actor, the race is confined by eliminating heats to six young watermen no more than a year 'out of their time', i.e., their apprenticeship. It is rowed under the auspices of the Fishmongers' Company from London Bridge to Chelsea. More than one winner of the colourful livery has gone on to attain international sculling fame.

Another sign of the times was the return of certain wartime anti-aircraft ships, minesweepers and transports to their proper duties as Thames 'Butterfly' ships, running summer day trips down to the sea.

Best of all was the revival of the river's traditional role as a setting for pageantry. We had had a reminder of what the Thames could do to celebrate victory; in 1951 it helped to give the war-weary nation a tonic by providing a background for flags and fireworks of the Festival of Britain. The Festival Hall on the South Bank was opened with much pomp by that most democratic of kings, George VI. In 1953 the river was very much *en fête* for the Coronation of Queen Elizabeth II. Many overseas visitors were taken round the Port, and the Queen of Tonga, the Ruler of Kuweit, the Sheikh of Bahrein and others brought a touch of colour to sombre dockland.

Liners, serving as hotels, berthed in the river and warships occupied commercial moorings while yachts of many nations tied up wherever the Harbour Master could find space for them. London school children were taken by water to places where they could see the Coronation procession. An older and more picturesque Thames made its contribution to the cavalcade: the Queen's Bargemaster and the Queen's Watermen. Their gorgeous uniforms and their honoured place in the pagentry were symbolic of the days when the river was the safest route for the conveyance of monarchs and their royal trappings.

Hundreds of thousands of spectators stood twenty deep along the Victoria Embankment to watch the splendidly extravagant firework display. Those of us who had endured the grim misery of six years of war were conscious of participating in a very great occasion and felt like echoing the words of Samuel Pepys similarly thrilled by the sights and sounds at the Coronation of Charles II: 'Now after all this I can say that, beside the pleasure of the sight of these glorious things I may now shut my eyes against any other objects, nor for the future trouble myself to see things of state and showe, as being sure never to see the like again in this world.'

But only a year later, in 1954, we were to see something nearly as exciting when the Royal Yacht *Britannia* came up the Thames with the Queen and Prince Philip at the end of their Commonwealth Tour. On this occasion I was afloat with the Harbour Master at dawn in the launch *Nore* which led the Royal Yacht and the escorting Trinity House *Patricia* up to the Pool, and then preceded the Royal Barge up to Westminster.

Once again the tideway was a riot of bunting interspersed with 'Welcome Home' messages, and the cheering kept pace with the progress of the yacht. We were then perhaps more sentimental and those of us who made that breath-catching passage up the river with the Queen were stunned by the emotion, the noise and the brilliance.

Every ship in the Port was dressed overall and all outward-bound vessels, even those belonging to countries not too friendly with us at the time, punctiliously dipped their ensigns as they passed *Britannia*. But it was the more homely sights that delighted me most; the two shabby little shrimping bawleys in the outer estuary dressed overall like any great liner but with miniature flags; the members of the canoe club who paddled out as a fleet from Canvey Island to acclaim their Sovereign; the little Gravesend wharf steam crane whose squeaking 'cock-a-doodle-do' on its whistle was answered by the booming voices of the P & O liners, *Arcadia* and *Himalaya*, across the river in Tilbury Docks; the precision with which the boys of the training vessel *Worcester* at Greenhithe manned ship to cheer their Royal Patron; the factory at Woolwich which morsed 'Welcome home' on its steam whistle. No one in those days could have questioned the loyalty of the People of the River.

A truer test of the loyalty of Thames-siders was the Royal Silver Jubilee celebrations in the summer of 1977. During the twenty-five momentous years of the Queen's reign, both the Thames and its people had changed. The river had been seeking to replace its former traditional trade with a vanished empire. And during that quarter-of-a-century the Port had moved downstream from those tidal reaches where it had flourished for nearly two thousand years. The People of the River, like the nation as a whole, had become less inclined to venerate tradition. But the resurgence of loyalty marking the Silver Jubilee lacked nothing in warmth and sincerity.

On 9 June, the old river became once again a royal road when the Queen progressed in the PLA launch *Nore* from Greenwich to Lambeth. But on this occasion she disembarked at selected points to meet her subjects and attend special celebrations organised by local authority and private citizens. At these more or less informal meetings affection was unrestrained and reflected the new spirit between throne and people. In the evening, a river pageant voyaged between Blackwall and Vauxhall; it consisted of three miles of ships, boats and floats organised and marshalled by the PLA with the co-operation of commercial firms and riverside institutions. Magic emerged on the return passage when the pageant was illuminated, when the tideway became a dream river of surpassing beauty.

There was more to come. At 10.30 p.m., a firework display began. It consisted of six sequences, each with a predominant theme of colour and effect, accompanied by a broadcast special musical programme. The spectacle increased in volume and complexity until, for the many thousands of watchers, it became a night of enchantment, far removed from and superseding social concerns of the day.

Although these and other great occasions on the Thames have stirred me profoundly, perhaps the most dramatic of all was the final stage of the State Funeral of Sir Winston Churchill on 30 January 1965.

The setting of his last journey on the river - and he had known its many moods in peace and war - was worthy of the man who by his spirit and example had saved this country, if not western civilisation. In the background were the grey-muzzled old Tower of London and the much earlier church of All Hallows, both peopled by the ghosts of many great actors in the English story. Upstream was the former London Bridge, soon to be replaced by the present structure. In the foreground was the Pool of London hiding its legends beneath a sober mercantile face.

The minute guns on Tower Wharf sent the gulls wheeling, and many of us thought of that other and more sinister gunfire which had called Churchill to leadership of the nation. The distant chords of Handel's Funeral March caused eyes to blink, and the lamenting pipe bands caught us by the throat as the cortège neared Tower Pier. At last came the bearer party with the coffin on which lay the insignia of the Most Noble Order of the Garter. It was carried aboard the PLA launch *Havengore*, and the family mourners filed into another PLA craft, *Thame*.

Lines were cast off and the Naval party piped him away as he had been piped so many times in two world wars. The well-rehearsed cranes along Hay's Wharf frontages on the south side bowed their jibs in unison. The two craft, led by the Trinity House *Landward*, moved off upstream and gradually disappeared in the winter haze.

'Long stood Sir Bedivere revolving many memories, till the hull look'd one black dot against the verge of dawn, and on the mere the wailing died away.'*

---

* Alfred Lord Tennyson, *Idylls of the King*.

chapter eighteen

End and Beginning

1

THIRD TIME LUCKY it seemed when, after the setbacks of two world wars, the finishing touches were put to the modernisation programme started by Lord Devonport* in 1909. By the end of the 1950s, Viscount Simon, PLA Chairman, and Sir Leslie Ford, General Manager, could claim that sheds, quays, warehouses, tugs and other craft, cargo-handling gear and techniques were at last up-to-date. But luck proved to be fickle, for many of these achievements were overtaken by the unforeseen container revolution and its swift repercussions.

Nevertheless, certain advances in the fifties prepared the way for things to come, notably the introduction of the Thames Navigation Service and the building of this country's first bulk-wine berth.

Electronics were already hard at work in the Port. The use of short-wave radio communication had spread from the dock police to Harbour Service launches and dock tugs, as well as to many river tugs of the towage companies. Such a link-up was invaluable to a port strung along sixty-nine miles of what was then the world's busiest waterway.

But PLA's Thames Navigation Service, started in June 1959, extended this network into the fields of radar and VHF radio telephony. It aimed at turning an omniscient eye on the ever changing river scene and giving navigational advice to ships. Dockmasters, wharfingers, towage firms, road transport and port labour could also be kept in the picture.

Since then, the service has been broadened to cover the whole of the

---

* See Appendix A - page 173 *et seq.*

Plate 21:

HM The Queen returns to Royal Yacht *Britannia* during the Royal River Progress, 9 June 1977. *Photo: Port of London Authority.*

Plate 22:

PLA Thames Navigation Service operation room, Gravesend, which, with the aid of short-wave radio and radar, controls the movements of all vessels entering or leaving the Port. *Photo: Port of London Authority.*

estuary and the tidal river up to Woolwich. Shore-based radar provides the duty officers in the Operations Room at Gravesend with a continuous display of these areas. The service now also has powers to direct the passage of all ships in port and co-ordinates some one thousand such shipping movements each week.

Reports on visibility, navigational hazards and other matters affecting ships in port converge on the Operations Room over a web of direct land lines. Automatic tide gauges with telemetered radio links give the duty officers the actual, as opposed to the predicted, depths - of some importance to a deep-draughted vessel for which a few more or less inches under her keel might mean all the difference between anchoring and waiting for the rising tide or moving up river right away.

This flood of information is co-ordinated in the Operations Room on a large-scale panoramic plan of the whole Port. It is virtually an illuminated picture, twenty feet long, which shows a constantly changing pattern of ships, dredgers, visibility, state of tide and so on. A viewing gallery allows pilots from the pilotage station next door to study what they are likely to meet when they bring in their next ship.

Vessels approaching the Port are given relevant details of the picture by VHF radio telephone and are directed to anchor or manoeuvre as the situation demands. Some very large tankers have almost literally been talked into their berths.

The Operations Room with its ripple of voices passing to distant ships, with its long illuminated chart, with its pulsating radar screens and with its undertones of professional assurance and authority, creates one of the most dramatic atmospheres in the Port. It epitomised for me the developments since my early days when shipping intelligence was supplied by Horace Martin, grandiosely designated 'The Shipping Clerk', who used to level his telescope at vessels passing his Gravesend balcony and then telephone the news to head office.

Also in 1959, the first bulk wine berth was built at the London Dock. The history of wine containers is to some extent the history of the trade in fermented grape juice. During Roman times wine came to Britain in earthenware amphorae and probably also in skins, although Herodotus (c. 484–425 BC) tells of it being transported in palmwood casks. By the time of the Plantaganets, France was sending wine to London in tuns, very big casks each holding 252 gallons. The trade was so important that from the word tun derives the modern space measurement, the register shipping ton of 100 cubic feet. And right up to the early 1950s, this country's imported wine came mainly in casks of varying sizes (of course, excluding such aristocrats as the chateau-bottled wines from famous vineyards).

With the rising costs of making and replacing casks, the wine trade, in the post-war years, had been experimenting with bulk containers which also meant lower transport and handling costs. For storage, PLA built a new berth at London Dock equipped with reinforced glass-lined concrete tanks holding in all some 200,000 gallons. It was modern; it was hygienic; it dispelled some of the Tolkien-like mystique of caverns and cobwebs hitherto associated with the trade.

The non-frothing electric pumps at the berth could discharge containers (brought alongside in barges) and, later, wine-tank ships, at some 9,000 gallons an hour; a river of wine not hitherto met with outside legendary celebrations. When duty had been paid and the wine was ready for delivery, it could be likewise pumped direct into road tankers.

The importance of the innovation to the Port was that, apart from oil, grain, sugar and other cargoes normally carried in bulk, it was virtually the first step in those modern techniques which are today rendering so many dockers redundant. Two hundred thousand gallons imported by traditional methods would need some 4,000 casks with high labour costs for all the manhandling involved. At the bulk wine berth, one man with two assistants controlled the whole operation by push button.

The scheme has proved so successful that a second berth has been built at the West India Docks and today the total storage capacity for bulk wine in the Port is 2 million gallons.*

2

In 1961 I saw the last stages of one of the most difficult salvage operations ever undertaken in the Port - the removal of what remained of the Great Nore Towers (one of the Maunsell Forts which had helped

---

* This is now the capacity of the West India Docks berth; the first bulk wine berth at the London Dock has been closed.

to defeat the aerial minelayers and the only one inside what was then the Port's seaward limit).

The Nore Fort - with the profile of a horror-film giant spider - consisted of seven towers, each formed of four heavily reinforced concrete legs supporting gun platforms, crew quarters, etc., connected by flying bridges. When the war ended, this fort, then known as the Great Nore Towers, was used by Trinity House for navigational light and sound signals in place of the *Nore* light vessel, withdrawn in 1943.

In 1953, the m.v. *Baalbek*, groping her way out of the fog-bound river, collided with one of the towers, causing it and another to collapse with the tragic loss of four of the fort's watchkeepers. At once the PLA asked the Government to remove the wreckage and what was left of the fort as likely to be a hazard to navigation.

On the principle that what the eye does not see, the heart does not grieve over, the Government merely cut off the tower legs and removed the superstructure. On the river bed remained seven hollow cruciform reinforced concrete bases composed of four arms, each approximately 82ft long, 7ft wide and 7ft deep. Each base weighed about 175 tons when sunk and had later been surrounded and covered by more than 1,000 tons of rubble.

For the Port this was no solution since the bases would prevent dredging to maintain the statutory depth at this point. Accordingly, in 1959, 'do it yourself' was decided on and the Port Salvage Officer began the task of lifting the bases.

Divers had reported that some of these had been fractured by the explosive charges used to cut the tower legs, allowing sand and mud to accumulate inside up to a weight of over 400 tons. Also, some of the arms were broken, but still joined by the reinforcements.

The powerful salvage vessel *Yantlet* and a fleet of salvage lighters were assembled. But work had to be intermittent because of the need for calm weather and suitable tides. It eventually took three summers to complete the job.

On one occasion the *Yantlet* and two lighters, aided by the rising tide, were slowly lifting a base out of its bed of rubble - the foredecks of the lighters awash and daylight showing under their sterns - when one of the lifting lugs on the base pulled away. The *Yantlet*'s bow shot out of the water and the two lighters suddenly took the full load and almost stood on their heads. Then their lug broke off and they, too, tried to take off.

Another time, five lighters dealing with one of the bases all had their bows under water – in one, it was almost lapping into the accommodation hatchway. Two of the older lighters, their hulls weakened by years of punishing work, wept at the strain and only constant pumping kept them afloat.

With these and somewhat similar misadventures, the work was finished in the summer of 1961. I watched the last base, slung beneath a staggering scrum of salvage craft, pass upstream to be beached amidst the triumphant blowing of steam whistles.

The divers had had the worst job, and the salvage officer in charge told me he always knew how things were going in the depths by what came up to him over the intercom. If all went according to plan, there would probably be snatches of a hummed refrain; if the reverse, the language was usually unprintable. One of the divers preparing for a long stay brought what he thought were his own provisions, only to find that he had instead a plentiful supply of tinned cat food!

The Thames produced its usual little surprise right at the end. Inside the last bit of loose wreckage – a steel catwalk – to be brought up were found a six-foot conger eel and a two-foot ragworm.

3

When I retired in 1962, although ships and some craft as well as dock cranes and warehouses had acquired different profiles, and some types of cargo had changed, and well-known landmarks and daily sights had gone, the Port was in many ways much the same as I had found it in 1918.

Some of the People of the River were more professional and many undoubtedly more independent and even bloody-minded, but they were bringing the same traditional skills to their daily work. Gravesend, Wapping and Rotherhithe were still the breeding grounds of families whose forbears for generations had worked in or around the same waters. Now, however, they used diesel engines, where their fathers had used steam and their grandfathers canvas or sculls; radar, where formerly only good eyesight and experience had kept craft out of trouble; radio telephones or loud hailers, where bunting or leather lungs had had to communicate over distance.

But one significant change was accelerating by the time I left the Port

- a decline in the warehousing trade. In the old days of long and uncertain ship passages and poor communication, merchants of this country had been glad to keep large stocks in bonded dock warehouses. After the war, an importer could telephone for an appointment on the other side of the world, be there in a few hours and buy his produce on the spot. And the faster and more reliable ships meant that he need no longer keep capital tied up in warehoused goods. With this decline have gone many of the unique processing and sampling services in the warehouses which I have described. (In the London World Trade Centre at the former St Katharine Docks, facilities for new trades include the assessing of products by specification instead of by samples.)

4

Twentieth-century western man is inclined to be somewhat cynical about Government committees. So, while we admired the incisive and perceptive conclusions of the Rochdale Report, when it was issued in 1962, few of us realised that here was a fuse leading to the abrupt disappearance of much that I have described.

At the beginning of the 1960s, there had been some highly vocal discontent amongst shipowners and traders about the state of British ports, causing the Government to set up this inquiry under Lord Rochdale. For London, it was the first searching look, uninhibited by pressure of vested interests, at the fragmented port administration, compromises and anomalies imposed when the Authority was created in 1908. The report contained criticism and praise as well as constructive and stimulating suggestions.

Rochdale stressed the need to introduce more cargo-handling machinery into our ports - something which all port authorities had been trying to do for many years. Now began a properly planned campaign to educate the London docker away from his Luddite tendencies. At old-style conventional berths, only limited success was achieved, but where berths were already mechanised men could see that by the use of machines they were already earning higher pay with less physical strain. Leaping forward a few years, agreement was eventually reached for the new container port at Tilbury to work round the clock every day of the year and for the introduction of a two-shift system at other berths. These measures meant that costly cargo-handling machines, hitherto idle for over two-thirds of their life, would now be considerably more productive.

Returning to the Rochdale Report, another welcome recommendation was that more deep-water berths were needed, and there was approval for plans already made to extend Tilbury Docks. In 1963, the foxes and the adders were dispossessed, the massive foundations of the wartime Pluto works demolished, and construction started on two new berths for deep-draughted conventional ships and two for drive-on vessels. More important, as was later apparent, these were so planned that the area could be developed further when necessary.

Another Rochdale recommendation led to the extension of the seaward port limit so as to take in the vital approach channels in the outer estuary which, in turn, led to the creation of the new Knock John Channel for deep ships (as I have already described).

Other plans to move more boldly into the computer age and to make fuller use of the 'tools of management' (work study, organisation and methods investigation, operational research, staff training schemes and management accountancy) gained momentum. It was all a far cry from my youth - from warnings about 'keeping the blee'n ledgers neat and tidy'.

The most far-reaching Rochdale recommendation, however, was the clearly defined need to end casual labour. This system of employment in the port industry had grown out of the fluctuating movements of ships, especially in the days of sail when a change of wind might delay arrivals and cause hardship, even actual starvation, to hundreds of dock workers and their families. From this had come the solidarity, bitterness and violence characterising most labour disputes in the industry - the 'one out, all out' attitude, and the brick through the window of the blackleg.

With the passing of sail, port employers gradually became more humane, and several attempts had been made to find a way of ending the practice of casual employment. But, although modern cargo liners were not so dependent upon weather and ran to a schedule, the hard fact remained - either there might be many ships in port requiring a large labour force for discharge and loading, or only a handful of vessels needing far fewer men.

By now pay and working conditions of dockers were among the best in comparable industries and 'fallback' rates were paid if no work were available. Then a more limited but concentrated inquiry under Lord

Devlin, aimed at averting a threatened dock strike, pointed the way to permanent employment. This was finally achieved in two stages. Henceforth piece work was abolished in favour of a standard wage, and the men had permanent employment.

The final swing of the pendulum was completed in 1972 with the Aldington* - Jones Report which gave all registered dockers in major British ports a substantial guaranteed wage for their working life, whether work was available or not, and a golden handshake (at the time of writing, up to £8,500) as an inducement to older dockers to leave the industry. It was hoped that this would speed up voluntary redundancy already begun. During the last ten years, the registered work force in the Port of London has been reduced from 24,000 to 8,000.

But all too often, it has been the younger men, able to take alternative employment, who have left, and the average age of dockers in the Port today is over forty-five years, and many of them, not yet of pensionable age, are too old or unfit to take on the full range of dock work.

All this meant that the docker had, in fact, virtually priced himself out of a job. The way was now open for the introduction of much more ambitious and much cheaper labour-saving methods which began to develop at a fantastic pace, some of them bordering on science fiction themes.

Even more far-reaching than the Rochdale Report was work, started in 1966, on a second and third stage of the Tilbury Docks extension, eventually to provide six container and three unit-load berths in addition to the first four berths already in use.

Each of these new-style berths needed some twelve to twenty acres of hinterland for the stacking and manoeuvring of containers. Tilbury, alone among London's docks, had plenty of undeveloped ground where the PLA was able to provide the first purpose-built container port in Britain.†

Even at the prices then current, the cost of establishing a container berth was enormous - some £2½ million - while it was also necessary to build a liner-train terminal and set aside a groupage area where

---

* Lord Aldington, Chairman of the Port of London Authority 1971/7.
† Begun by Mr Dudley Perkins who had succeeded Sir Leslie Ford on his retirement as chief executive.

containers could be 'stuffed' or 'stripped' as required, and to provide a fleet of fork-lift and straddle-carrying machines.

This huge outlay meant that the berths must show a very high rate of productivity. In effect, they can each handle more than a million tons of cargo per year as compared with some 100,000 tons by conventional means. To a shipowner, saving in time is money. Every moment spent by a vessel as a floating warehouse incurring dues in port instead of earning her keep by carrying cargo across the sea is a dead loss. Ships at the Tilbury Container Port are now turned round in a matter of hours instead of, as formerly, weeks.*

The unit-load berths handle packaged timber, i.e., timber strapped together by steel bands and, in effect, a single unit. A berth crew can handle as much as 4,000 tons of such timber in a ten-hour working day. These berths effectively put out of business the remaining timber porters, the post-war timber cranes and what were described as the finest timber storage sheds in the world at the Surrey Docks.

Another form of unit-load handling was successfully introduced at Millwall Dock with a new terminal for the London-Canary Islands service. Purpose-built ships have hydraulic lifts and side doors to allow loading and discharging to be carried out by fork-lift trucks and mobile cranes. A full load of some 2,000 tons of cargo on pallets can be dealt with in one day.

Still the search for labour-saving technology goes on, each invention posing new threats to the old-style docker. Ships now come in carrying their own ready-loaded barges which are lifted from or lowered to the water by the mother ship's elevator. Yet another type is a multi-purpose drive-on vessel, handling cargo on pallets, in containers, in unit loads and on wheels, the ship equipped with her own fork-lift trucks and tractors.

Not surprisingly, some dockers, despite high wages and permanent employment, have been alarmed. There have been ugly cases of violent picketing at private container depots and warehouses, the militants demanding that the work of 'stuffing' and 'stripping' containers should be done by dockers, irrespective of the distance from ports. 'Blacking'

---

* A riverside container terminal in Northfleet Hope Reach, outside Tilbury Docks, has recently been completed and is now in use.

Plate 23:

Part of the container terminal, Tilbury Docks, showing shipping at the multi-user berths. The in-dock end of the new riverside Northfleet Hope berths can be seen top right. *Photo: Port of London Authority.*

Plate 24:

Now a major London attraction, HMS *Belfast* at her permanent moorings in the Pool of London.

of lorries has sometimes been accompanied by deplorable scenes, as if the men were back in the bad old days when the next meal depended upon getting work.

But the days of mere muscle power in the loading and discharging of ships are probably numbered. Before many years have passed, it seems inevitable that individual highly skilled and highly paid dock workers operating ingenious machines will have replaced the traditional gangs – and the methods of the Phoenicians will have gone for ever.

To provide for ever larger bulk-grain ships, too large and drawing too much water to go up to the enclosed docks, the PLA built a new grain terminal on the riverside at Tilbury. Hitherto, these giants had discharged at Continental ports, grain for London being transhipped. The new terminal has already brought much of the direct trade back to London. Ships of over 80,000 dead weight tonnage can be handled; all working is automatic and aided by closed-circuit television. Some 20,000 tons have been discharged in a fourteen-hour working day of two shifts, and the berth has storage capacity for 100,000 tons. The grain trade has come a long way since I used to admire the shadowy pattern of ropes and spars cast by the old windjammers on the mills' walls at the south side of the Royal Victoria Dock.

5

This wave of labour-saving and changing methods throughout the Port has had many repercussions. Although some of the big wharves down river still provide heartening success stories, others, hard hit by the decline in warehousing, by the growth of container ships and bulk carriers which can use only the lower, deeper reaches, and by the high wages of dockers arbitrarily allocated to them for permanent employment, have gone out of business. The Pool of London, where formerly ocean-going vessels berthed at New Fresh Wharf and the complex of Hay's Wharf, now sees only small craft and an occasional visiting warship. (HMS *Belfast*, at permanent moorings in the Pool, looks strangely out of place.)

Indeed, from the Pool downstream there are depressing lines of empty warehouses, mouldering quays and idle cranes awaiting planning decisions. In the background are the closed-down basins of the former

London and Surrey Commercial Docks. A number of berths has also been closed elsewhere in the no longer viable West India, Millwall and Royal Docks.* Other significant changes include the sale by the PLA of its vast Cutler Street Warehouses in the City of London and its stately office pile on Tower Hill. Devolution has been substituted for central-isation, and severance payments and retirements on pension have sub-stantially reduced the size of the Authority's staff in all grades.

So far as traffic on the river is concerned, at least in the middle and upper tidal reaches, the picture is equally depressing to an old hand. Most of the sleek passenger liners have gone, and the giants of the Port are now found only in the lower reaches. In size they are more impressive than the old-time ships, but many of them look somewhat like aircraft carriers or floating docks and less and less like the vessels to which tradition accustomed us.

The General Steam 'Butterfly' steamers have gone. Now that many dockland families tend to holiday on the Costa Brava or in the Canaries, they are less interested in a river trip to Margate or Clacton. Both hydrofoil services and hovercraft have represented the river's wind of change but neither is evocative of the one-time Cockney outing.

The traditional lighterage trade, almost as old as the Port itself, no longer monopolises each tide with its teeming craft. The trade has suffered badly from the decline in warehousing (i.e., water transport to and from dock or wharf storage) and from competition by containers and road transport. Whereas some 4,000 lightermen were formerly employed on the Thames, the total now is little more than 1,000. And more than one hundred lighterage firms have gone out of business during the last ten years. Pressure is, however, growing for the Thames and its linked navigable waterways to be used again as a major transport artery; this has the support of dedicated environmentalists. And industry is beginning to remember that it is cheaper to transport goods by water than by land.

Controversy over the proposed seaport at Maplin (at first linked to the projected airport) has temporarily died down, but the subject is by no

---

* See Appendix B.

means dead. Supporters of the scheme claim that without such a new deepwater port able to accommodate the largest ships using the English Channel, the giant Continental ports will eventually capture much of our overseas trade, London then becoming merely a transhipment terminal.

The scheme would profoundly affect the present Port and might result at some future date in London's seaborne commerce being handled almost exclusively in the estuary.

The effect of these and other changes likely to come will be incalculable. The tidal river is bound to have a different role from that which it has played for some two thousand years. Conferences, official inquiries, discussions and countless unofficial speeches have all pointed to the need for a new Thames-side, most of them visualising a tidal river of pleasure, bordered by green vistas, modern homes, new industries and offices, instead of a shabby, timeless shopping street of nations.

Whatever the role of the tideway in the future, we must not forget the past, for the river's tale is a microcosm of our island story. Patriotism in this country is not entirely dead, and many of us feel that our creditors (and detractors) in the Old World and the New need to be reminded that much of the blood and treasure going out on Thames tides through the ages has never been repaid; that some nations which now patronise us owe their very existence to us.

For my part, I must be content with memories of my last journey upstream into a sunset when dockland's *chevaux de frise* of crane jibs, conveyors, masts and funnels appeared like a sombre etching against a Turneresque smoke-blue and crimson arch. And that arch seemed a sort of proscenium for kings and queens and statesmen, architects, painters and poets, scientists, engineers, seamen and merchants, whose imagination, brave hearts, keen minds and inspired hands through the centuries have helped to make this proud coming-together of ships, commerce, building and tradition; this unique Spirit of London's River.

'Sweete Themmes! runne softly till I end my song.'*

---

* Edmund Spenser, *Prothalamion.*

APPENDIXES

Appendix A

Historical Background

1

TO UNDERSTAND the basic cause of the Port of London's condition when I entered the PLA service in 1918, one must look back to the early nineteenth-century dock-building boom.* Briefly, enclosed docks linked to the river by locks were the answer to two big problems: slow suffocation of the river by the rising volume of trade, rooted in Elizabethan expansion; and cargo plundering from ships and barges on a scale which inflicted huge losses on importers, owners of riverside wharves and warehouses, and the Crown. In the enclosed docks, ships could lie alongside in tide-free water while their cargoes were protected by massive walls guarded day and night by the dock companies' own police.

But towards the end of the last century, savage competition, with excessive rate-cutting, between the surviving dock companies, had produced another crisis in the Port. Accommodation for ships and cargo had become inadequate, and so had the tidal Thames, then controlled by the Thames Conservancy Board.

A Royal Commission, which reported in 1902, advised the creation of a new authority, representing all users of the Port, to take over the dock companies' assets (London and St Katharine Docks at Wapping; East and West India and Millwall Docks at Poplar; Royal Albert and Royal Victoria Docks in Newham; Tilbury Docks in Essex; and Surrey Commercial Docks at Rotherhithe; together with large areas of unde-

---

* A full account of the early history of the Port of London, including the construction of these docks and the founding of the Port of London Authority, is contained in Sir Joseph Broodbank's official *History of the Port of London*. (Two volumes. Daniel O'Connor, 1921.) See also *The Port of London* by R. Douglas Brown (Dalton 1978).

veloped riverside land) and conservancy duties over the whole of the tidal Thames - that is, between Teddington and the sea.

The sweeping powers recommended by the Report were, however, attacked and gradually reduced by entrenched opponents; when the Act creating the Port of London Authority was eventually passed in 1908, the originally bold and imaginative scheme had become the inevitable compromise.

The new Authority was to own and operate the enclosed docks, although the custom of the Port allowed private employers to function there as stevedores, the PLA being only one of too large a number of employers of dock labour. Thus many activities carried out on PLA premises were completely outside its control.

On the river, the PLA as conservators had no part in pilotage, navigation lights and marks, port health, river police, towage and lighterage, all of which remained in other hands. The riverside wharves, still in private ownership and many of them rivals of the Authority in the warehousing trade, continued to retain the traditional right to send barges into the docks to collect or deliver cargo from or to ships there without payment of dues. This right, known as 'The Freewater Clause', had been inherited from the era of the dock companies whose attempt to annul it had led to the Royal Commission and so to the creation of the Authority. The riverside wharfingers were, moreover, strongly represented on the PLA Board (consisting of twenty-eight members [later reduced] partly appointed and partly elected by users of the Port). Sir Joseph Broodbank, a member of that first Board, made the astringent comment: 'With such a constitution the word "Authority" is a misnomer; no authority in England can have less freedom and more limitations.'

Docks and equipment urgently needed to be repaired or replaced. The dredged channel in the river (belatedly started by the Thames Conservancy before it handed over to the PLA) had to be developed to a statutory width and depth laid down in the Act.

2

The Authority's first Chairman, Lord Devonport, was the organising genius who tackled, and to a limited extent solved, this gigantic jigsaw. Although to all port workers he was something of an ogre, he was in many ways the right man for the job. He was outside

Plate 25:
A good catch during an angling competition at Gravesend. Thirty years ago, the tidal river was almost completely devoid of fish: today there are nearly 100 varieties of fish in the lower Thames. *Port of London Authority*

Plate 26:

Viscount Davenport, PC, the first PLA chairman. Portrait by Z. Nemethy, after de László. *Photo: Port of London Authority.*

the conflicting interests of the Port, and his relentless drive and fierce impatience were probably what were needed to restore the Thames and its docks to vigorous commercial life.

He was one of the last of the Victorian masters who believed in rule by fear. The appearance of this impressive frock-coated figure at the docks or at the Authority's head offices kept everyone on his toes and perhaps, sad to say, produced better results, at least in the context of his time, than the more humane approach of today.

But our generation still suffers from the reverberations of those autocratic methods. So far as PLA executives were concerned, he was responsible for a slavishly conformist attitude towards management, and a stifling of initiative and critical appraisement, which persisted long after he had gone. Such submissiveness proved sterile, while the long-term effect on dock labour has been disastrous.

One story which has passed into PLA annals was typical of his ability to drive a hard bargain. During a visit to dock warehouses he decided to buy some choice carpets stored there. The owner, an Armenian Jew, was summoned to the Chairman's private room. Presently the latter's angry voice and the banging of a desk were heard by fascinated listeners. When the merchant finally emerged, pale and shaken, his reply to the question of whether Lord Devonport had bought the carpets was: 'Yes. Lord Devonport now owns them but I don't know whether I am to receive a cheque or give him one.'

In 1911 the dockers asked for a revision of pay and conditions which had not been amended since 1889. Impatient to press on with his plans, the Chairman (together with other London Port employers) granted pay increases totalling £200,000 a year and a few other concessions.

To the men this was a case of too little, too late, and the following year smouldering resentment flamed into a full-scale dock strike. Grimly Lord Devonport began to fight. Unmoved by Government pressure and public opinion, he coldly proclaimed that he would starve the workers into submission.

Towards the end of the strike, the men's bitterness overflowed and one of their leaders, Ben Tillett, led a vast crowd of dockers in prayer at a meeting on Tower Hill. 'Oh, God,' they intoned, 'strike Lord Devonport dead.' Their prayer was unanswered. After ten weeks of chaos in the Port, the strike was smashed and the work of development was resumed.

The great tideway dredging programme, the construction of the King

George V Dock at North Woolwich and the new head offices in the City at Trinity Square were completed under the Chairman's dynamic drive. But however fast the pace, Devonport cried 'Faster'. The New Entrance Lock, the New Dry Dock and the Passenger Landing Stage at Tilbury were all started during his regime. And these and many other improvements set a measure for the scope and pace of later developments.

Such was the man who was fighting to rejuvenate the shabby Port of London when I joined it in 1918; who cast a shadow of fear over most of the staff; whose paternalism persisted for more than a decade after his retirement in 1925.

When he died in 1934, the obituary notices were almost unanimous in judging him by his achievements rather than by his methods.

The portrait of Lord Devonport by de László in the former PLA head offices was a good likeness, reflecting something of his intolerance and pride, and also of his strength.

Appendix B

Crisis in the Port

1

ON 4 MAY, 1978, Sir John Cuckney, who in October 1977, had succeeded Lord Aldington as Chairman of the Port of London Authority, issued a remarkable paper setting out details of a major crisis in the Port.

This paper stated bluntly that the Port of London was on the verge of bankruptcy. For some years it had been failing to cover operational costs, and measures taken had been on the whole too little and too late. Spending accumulated reserves, closing some of the upper docks, selling surplus land (at a fraction of its worth by the fall in property values, the Community Land Act and the Development Land Tax) and borrowing were palliatives which no longer sufficed.

The paper emphasised that in addition to new thinking, new marketing methods and better co-operation from trade unions, a slimmer port would be essential. The retention of employees for whom there was no possibility of work, and of old and unfit men, would have to be faced. However, within the compass of this book, the most drastic proposal was to close the remaining upper docks - the West India/Millwall and the Royal Docks. Both groups had suffered severely from the steady decline in conventional cargo caused by the container trade, and they were no longer viable.

But the closing of these docks would create huge unemployment problems in London's dockland, and trade unions and local councils protested violently. The matter was submitted to the Government for consideration.

At Government's request, the PLA put up a modified plan, suggesting the closure of only the Royal group with a reduction of some 2,000 jobs. But Government rejected the closure of any docks. Realising, however, that the crux of the problem was surplus man power, the Secretary of

State for Transport promised financial aid if the **PLA** and the unions agreed specific targets for labour force reductions and more economical working.

A joint committee was immediately set up between the Authority and the unions (an innovation likely to be of great value in future port working) and discussions to reduce man power and costs and improve profitability began.

At the time of writing, negotiations are still in progress and it is hoped that agreement will at least allow the Port to overcome immediate problems. (These were further aggravated by the lorry drivers' strike at the beginning of 1979, costing the **PLA** approximately £1 million in lost revenue each week while it lasted.) Forward planning for the next five years is also being undertaken, and the deferred Maplin project is being kept under review.

The people of the river, like the nation as a whole, are punch drunk with change and the accompanying economic gales. But in Thames's long story much the same has happened before; and bold and adventurous leadership has repeatedly overcome crises. If the tideway is true to its great past, the future need not be in doubt.

INDEX